# Advances in Energy Recovery and Efficiency Technologies

*Edited by Basel I. Abed Ismail*

Published in London, United Kingdom

Advances in Energy Recovery and Efficiency Technologies
http://dx.doi.org/10.5772/intechopen.1003402
Edited by Basel I. Abed Ismail

Contributors
Abdelmajid Abouloifa, Basel I. Abed Ismail, Celal Yelgel, Chaouqi Aouadi, Ibtissam Lachkar, Jinela González, Katherine Chung-Camargo, Kent B. Hytken, Lei Guan, Meriem Aourir, Miguel Chen Austin, Mudasar Zahoor, Olga Yuil, Shaofeng Zhang, Shuo Chen, Sijing Guo, Tesfahun Meshesha, Thasnee Solano, Vivian Velarde, Yongle Chen, Övgü Ceyda Yelgel



First published in London, United Kingdom, 2025 by IntechOpen
IntechOpen is the global imprint of INTECHOPEN LIMITED, registered in England and Wales, registration number: 11086078, 167-169 Great Portland Street, London, W1W 5PF, United Kingdom

For EU product safety concerns: IN TECH d.o.o., Prolaz Marije Krucifikse Kozulić 3, 51000 Rijeka, Croatia, info@intechopen.com or visit our website at intechopen.com.

British Library Cataloguing-in-Publication Data
A catalogue record for this book is available from the British Library

Advances in Energy Recovery and Efficiency Technologies
Edited by Basel I. Abed Ismail
p. cm.
Print ISBN 978-1-83769-772-4
Online ISBN 978-1-83769-771-7
eBook (PDF) ISBN 978-1-83769-773-1

If disposing of this product, please recycle the paper responsibly.

# Meet the editor

Dr. Basel I. Abed Ismail is a successful, long-track, experienced academician and Associate Professor and Chair of the Department of Mechanical and Mechatronics Engineering at Lakehead University, a pioneering and international academic researcher in alternative energy and renewable and green energy engineering technologies. He is an academic editor for several books, an academic author for many peer-reviewed articles and book chapters, a licensed Professional Engineer (P. Eng.) of Ontario, Canada, and a member of the Canadian Society for Mechanical Engineering (CSME).

# Contents

# Preface

Energy recovery and efficiency technologies and methods are aimed at decreasing energy wasted or unutilized in various energy conversion processes, ultimately leading to a decrease in the required energy input (fuel) and improving the overall energy efficiency of systems and energy conservation of natural energy resources. These technologies, once efficiently designed and their performances are optimized, will potentially improve the economics and sustainability of systems, as well as reduce their emissions, air pollution, and thermal energy waste discharged to the environment. This book, titled "Advances in Energy Recovery and Efficiency Technologies", presents and covers unique topics related to advances and innovations in energy recovery & energy efficiency technologies. It is the result of contributions from several researchers and experts worldwide. An introductory chapter presented in this book will discuss low-grade waste heat energy recovery using thermoelectric power generation (TEG) green technologies. Also, a perspective chapter presenting some fundamental energy conversion aspects and modelling related to solar PV power generation with illustrative numerical case studies is included in this book. Other interesting chapters in this book include: downhole steam generation for heavy oil and oil sands production; Ocean energy harvesting history and technologies; Energy-efficient deep learning training; Grid-connected PV systems with filtering and reactive power injection; Energy-efficiency measures to achieve zero energy buildings in tropical and humid climates; Energy conservation measures in zero-energy building; the role of machine learning methods for renewable energy forecasting. It is hoped that the book will become a useful source of information and a basis for extended research for researchers, academicians, policymakers, and practitioners in the area of energy recovery and energy efficiency technologies.

The academic editor of this book would like to sincerely thank all chapters' authors and experts for their great efforts and the quality of the chapters presented. He would also like to thank Ms. Iva Horvat, the book publishing process manager at IntechOpen Science, for her excellent efforts in managing the publication process of this book.

**Dr. Basel I. Abed Ismail, P.Eng.**
Associate Professors and Chair,
Department of Mechanical and Mechatronics Engineering,
Lakehead University,
Thunder Bay, Ontario, Canada

Chapter 1

# Introductory Chapter: Energy Recovery of Low-Grade Waste Heat for Green Power Generation Using TEG Technologies

*Basel I. Abed Ismail*

## 1. Introduction

This book presents unique research topics of advances in energy recovery and efficiency technologies. Energy recovery and efficiency engineering refers to thermal or mechanical energy technologies or methods that aim to decrease or minimize the energy consumption or energy input of/to a particular system by the exchange of energy between a sub-system and the main system. The main goals of energy recovery technologies are to improve the overall system energy efficiency and its cost and sustainability, as well as to reduce its emissions, pollution, and waste energy to the environment. This book presents and discusses unique topics related to advances and innovations in energy recovery & efficiency technologies. It is aimed that this book will be of excellent value and provide state-of-the-art information for researchers, academicians, policymakers, designers, and engineering practitioners and technologists in the area of energy recovery and energy efficiency in various technologies.

Energy recovery is a method or a technology that could be used to capture and reuse wasted energy from potential sub-systems, thus improving the overall efficiency by minimizing the input energy of the main system. **Figure 1** shows classification of energy recovery methods into four categories [1]. Direct energy conversion and recovery from low-grade waste heat to electrical power using thermoelectric power generation is considered an active type and promising green technology [2, 3]. The global demands on electrical energy have been steadily increasing in the past few decades and are expected to continue to significantly increase in the future. However, when converting conventional (fossil fuels) or alternative energy sources into electrical energy, most energy generation technologies inevitably produce low-grade waste heat as a byproduct that is discharged to the environment, most of it at temperatures that are too low to be recovered by conventional electrical power generators [3, 4]. According to Ref. [2], waste heat recovery technologies can be classified into active and passive technologies as shown in **Figure 2**.

It was reported that each year more than 60% of the globally consumed energy as primary energy is lost during energy conversion, most of which is wasted as thermal energy (heat) released to the environment, as shown in **Figure 3** [5].

 IntechOpen

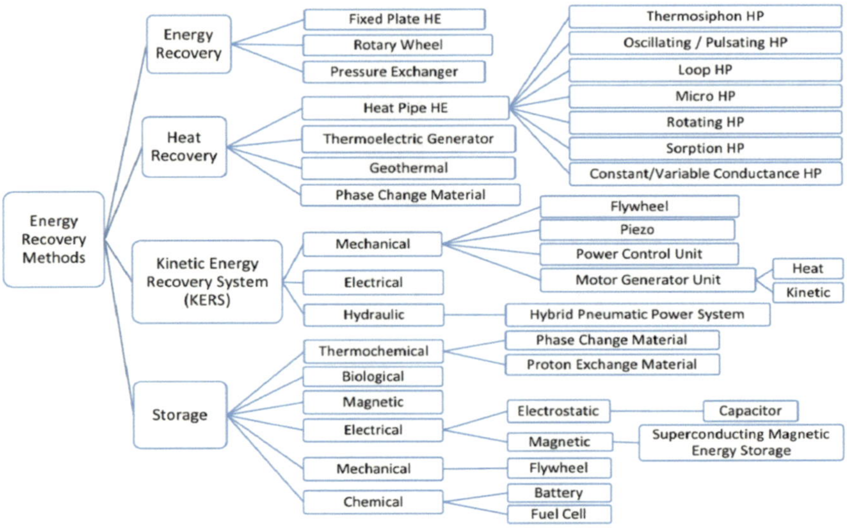


**Figure 1.**
*Classification of Energy Recovery Methods [1].*

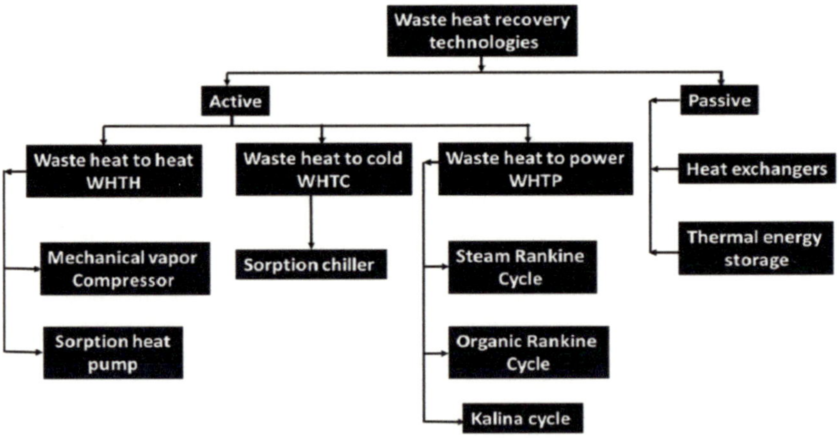


**Figure 2.**
*Classification of Heat Recovery technologies [2].*

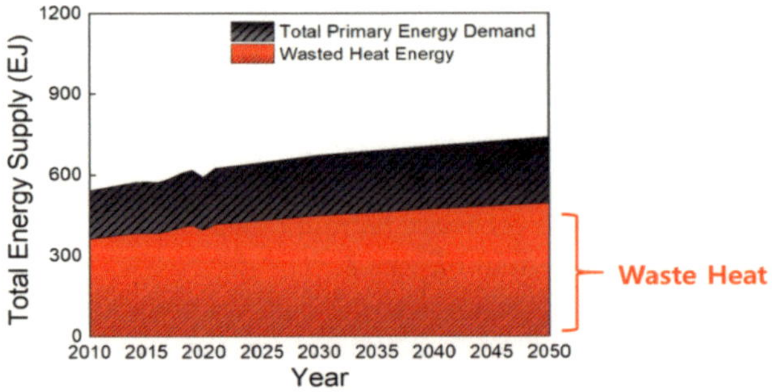


**Figure 3.**
*Projected 2050 trends of the global primary energy demand and its related wasted heat energy [5].*

Finding dependable, clean, and sustainable sources of electricity (primary energy) is one of the biggest challenges of the modern era due to (1) the global increasing demand on electrical energy, (2) conservation of natural energy resources, and (3) the ongoing stringent regulations of GHG emissions.

## 2. Principle of thermoelectric power generation (TEG)

Thermoelectric power generators (TEGs) are solid-state technology that works based on the principle known as *Seebeck effect*. **Figure 4** shows the theoretical principle of a simple thermoelectric power generator [3].

In this concept, there are two junctions, namely, the hot and cold junctions with an established temperature difference. These junctions are connected via two dissimilar metals or semiconductors (A & B), which as a result generate a voltage using Seebeck effect, as shown in **Figure 4**. This simple concept is used in thermocouples in applications involving temperature measurements. Therefore, TEGs can work as power generators. In **Figure 4**, heat $Q_H$ is transferred along materials A & B from the high temperature source, maintained at $T_H$ to the low temperature heat sink, maintained at $T_L$, where heat is rejected as $Q_L$. This forms a thermodynamic power cycle once it is connected to an electric load circuit. Using the first law of thermodynamics, the generated electrical power $W_e$ is simply the difference between the two heat transfer rates $Q_H$ and $Q_L$. **Figure 5** shows a schematic diagram of an actual design of a basic TEG device [3]. The TEG device consists of two ceramic plates with thermoelements sandwiched in between them. These thermoelements constitute $n$-type and $p$-type semiconductors. The thermoelectric performance of the semiconductors used in TEG devices can be evaluated based on a dimensionless figure of merit $ZT$. More theoretical details about the performance and advances of materials of TEG devices can be found in Ref. [3–11].

## 3. Low-grade waste heat recovery applications

Waste heat is normally from two sources, namely, (1) equipment energy conversion inefficiencies resulting in heat losses that could be prevented, and (2) heat losses due to the inevitable thermodynamics limitations governing an energy

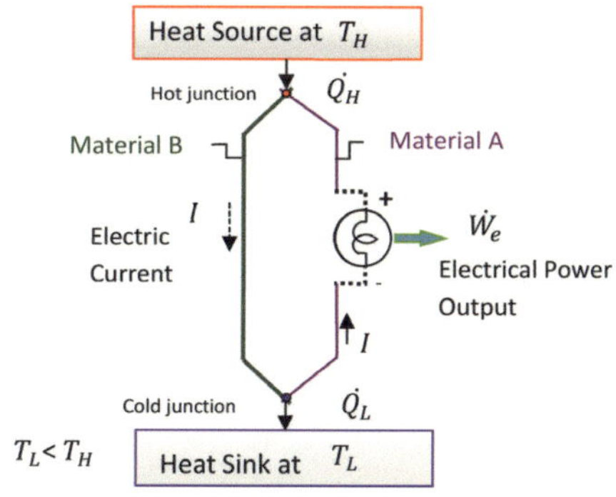


**Figure 4.**
*A schematic diagram showing the principle of a simple TEG [3].*

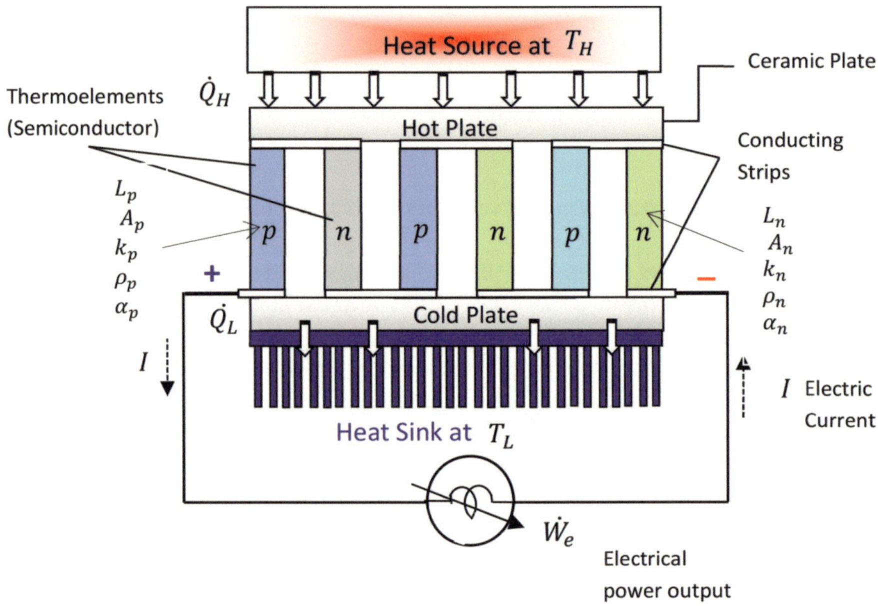


**Figure 5.**
*A schematic diagram showing the arrangement of a basic TEG device [3].*

conversion process that could be recovered and reused in another energy-consuming process [12]. Waste heat from various sources can be multi-grade [13]. **Figure 6** shows the three categories of multi-grade waste heat, depending on the level of the source temperature, in industrial applications including transportation applications [13]. The value and potential to reuse/recover waste heat depend on its temperature. Ref. [11] reported that low-grade waste heat recovery technologies are currently

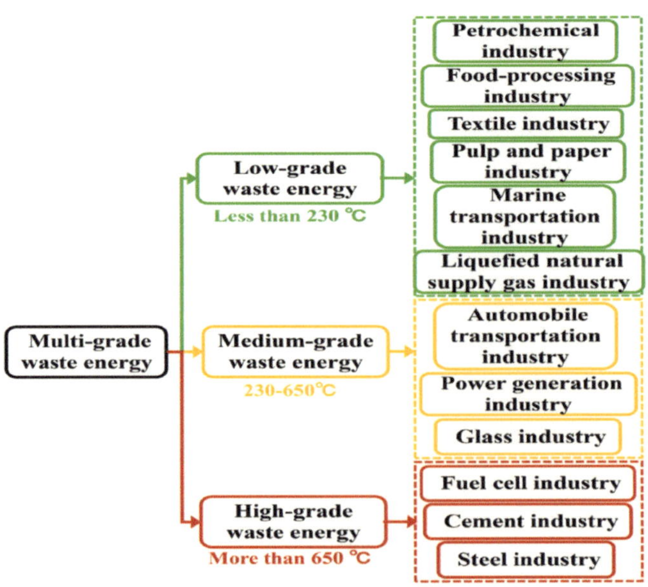


**Figure 6.**
*Classification of industrial heat wastes including transportation [13].*

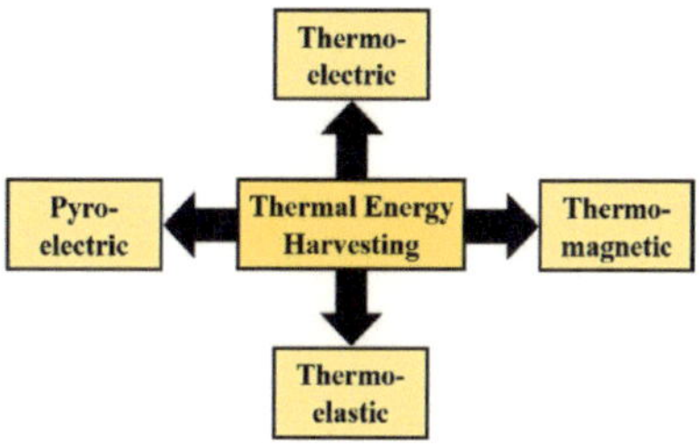


**Figure 7.**
*Current harvesting technologies of low-grade waste heat sources [11].*

dependent on four energy harvesting technologies, namely, (1) thermoelectric, (2) thermoelastic, (3) thermomagnetic, and (4) pyroelectric, as shown in **Figure 7**.

Low-grade waste heat recovery using thermoelectric power generation technologies has been extensively studied due to its merits and advances of thermoelectric materials and semiconductors. More details and discussions about low-grade waste heat recovery using thermoelectric power generation can be found in Ref. [9–13].

## Author details

Basel I. Abed Ismail
Department of Mechanical and Mechatronics Engineering, Lakehead University, Canada

*Address all correspondence to: baseliai@gmail.com

## References

[1] Aridi R et al. A comprehensive review on hybrid heat recovery systems: Classifications, applications, pros and cons, and new systems. Renewable and Sustainable Energy Reviews. 2022;**167**(112669):1-16

[2] Inayat A. Current Progress of process integration for waste heat recovery in steel and iron industries. Fuel. 2023;**338**(127237):1-16

[3] Ismail BI, Ahmed WH. Thermoelectric power generation using waste-heat energy as an alternative green technology. Recent Patents on Electrical Engineering. 2009;**2**:27-39

[4] Ismail BI. Power generation using Nonconventional Renewable Geothermal & Alternative Clean Energy Technologies. In: Carayannis EG, editor. Planet Earth 2011 – Global Warming Challenges and Opportunities for Policy and Practice. London, UK, Rijeka, Croatia: InTech Open Access Publishing Company; 2011 eBook ISBN: 978-953-307-733-8

[5] Hur S et al. Low-grade waste heat recovery scenarios: Pyroelectric, thermomagnetic, and Thermogalvanic thermal energy harvesting. Nao Energy. 2023;**114**(108596):1-22

[6] Ismail BIA, Abed JHI. Fundamental aspects and advances in thermoelectric materials for power generation: A numerical case study. In: Abed Ismail BI, editor. New Materials and Devices for Thermoelectric Power Generation. London, UK, Rijeka, Croatia: InTech Open Access Publishing Company; 2023 eBook ISBN: 978-1-83769-509-6

[7] Ider J et al. Efficiency analysis of thermoelectric generators. Materials Science and Engineering. 2024;**B 300**(117122):1-7

[8] Ma Q et al. Theoretical analysis and design optimization of thermoelectric generator. Applied Thermal Engineering. 2017;**127**:758-764

[9] Jaziri N et al. A comprehensive review of thermoelectric generators: Technologies and common applications. Energy Reports. 2020;**6**:264-287

[10] He J et al. Advances in the applications of thermoelectric generators. Applied Thermal Engineering. 2024;**236**(121813):1-39

[11] Safaei B et al. State-of-the-art review of energy harvesting applications by using thermoelectric generators. Mechanics of Advanced Materials and Structures. 2023;**31**:1-33

[12] Tohidi F et al. Thermoelectric generators: A comprehensive review of characteristics and applications. Applied Thermal Engineering. 2022;**201**(117793):1-44

[13] Su Z et al. Opportunities and strategies for multigrade waste heat utilization in various industries: A recent review. Energy Conversion and Management. 2021;**229**(113769):1-30

**Chapter 2**

# Perspective Chapter: Fundamental Energy Conversion Aspects and Modeling of Solar Photovoltaic (PV) Power Generators with Illustrative Numerical Case Studies

*Basel I. Abed Ismail*

## Abstract

Power generation using conventional and historically well-established renewable energy sources are well known for saving natural energy resources (e.g., fossil fuels), reducing GHG emissions and air pollution, as well as meeting future demands for prime electrical energy in various sectors and applications worldwide. For many years, solar photovoltaic (PV) has proven and continued to be successful and promising source of renewable energy for power generation. In this chapter, some fundamental aspects of power generation using solar PV systems are presented and discussed. Also, illustrative numerical examples based on fundamental energy conversion aspects and modeling are discussed with applications.

**Keywords:** power generation, solar PV, energy efficiency, battery system, maximum power

## 1. Introduction

### 1.1 Solar PV power generation

Photovoltaic (PV) energy converters are semiconductor devices that convert part of the incident solar radiation (in form light) into electrical energy. For this case, incident solar radiation can be considered as discrete "energy units" called photons. Photons have zero mass with zero charge. The speed of light $C$ (m/s) is governed by the following Eq. (1) in terms of the radiation frequency $\nu$ ($s^{-1}$ or Hertz, HZ) and its wavelength $\lambda$ (m) [1, 2]:

$$C = \lambda\,\nu \tag{1}$$

The energy of a single photon, $E$ (J), is related to $\nu$, using the famous Planck's Equation, given by (1):

$$E = h\,\nu \tag{2}$$

Where $h$ is Planck's constant = 6.6256 x $10^{-34}$ J s. Therefore, based on this equation, the most energetic photons are those with high $\nu$ (or low $\lambda$).

### 1.2 Principles of a PV cell and types of PV system

The most common PV cells are made of single-crystal silicon (Si). An atom of silicon in the crystal lattice absorbs a photon in the incident solar radiation (sunlight), and if the energy of the photon is high enough, an electron from the outer shell of the atom is liberated. Solar PV cells typically consist of two types of semiconductor material, often the *p*-type (positive charges, holes) and *n*-type (negative charges, electrons). In this atom ionization process, an internal electrical field (potential difference) is generated at the p-n junction. The electrons are swept away by passing through an external electric circuit outside the semiconductor materials. This process is shown in **Figure 1**. Contacts are made by metal grids on the top layer (uncovered for photons penetration) and metal bases on the bottom of the cell. There are many variations on cell material, design, and methods of manufacturing them. Some examples of semiconductor materials used for PV cells include cadmium sulfide (CdS), polycrystalline silicon (Si), and gallium arsenide (GaAs). The power generated by a PV cell is determined by the amount of incident solar radiation (sunlight photons), the type and area of the semiconductor, and the wavelength of the sunlight. There is a maximum $\lambda$ (thus minimum energy level) for photons that can generate an *electron–hole* (*p-n*) pair. It should be noted that solar radiation at higher $\lambda$ does not produce *electron–hole* pairs but rather heats the PV cell. In this case, the incident solar energy is converted into heat (cell internal energy losses).

## 2. Modeling and performance aspects of a PV power generator

The evaluation and optimization of the design and operation of solar PV systems are normally based on determination of the electrical characteristics (i.e., the I-V relationship) of the PV cells under various levels of (a) solar radiation and (b) PV cell temperatures. The DC power generated by a solar PV cell, module, or panel, can be analyzed and evaluated using an equivalent PV circuit shown in **Figure 2**. At fixed

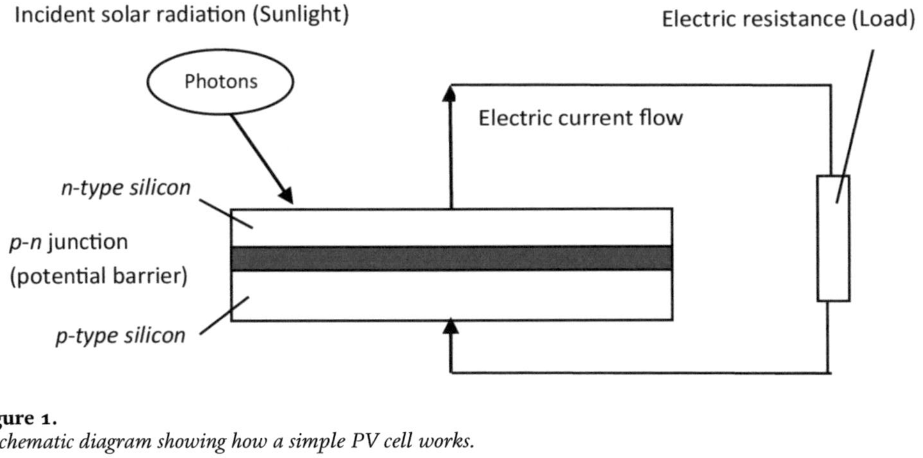


**Figure 1.**
*A schematic diagram showing how a simple PV cell works.*

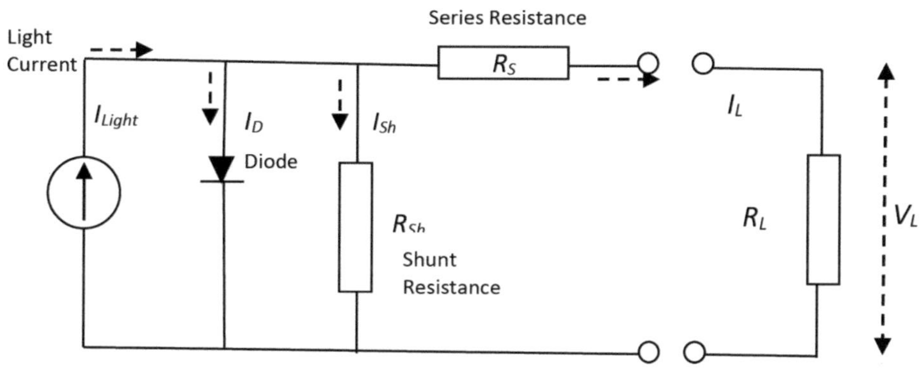


**Figure 2.**
*A schematic diagram showing an equivalent circuit of a PV cell, Adapted from [2, 3].*

| PV operation parameter | Definition |
|---|---|
| $I_{Light}$ | Light current, $A$ |
| $I_D$ | Diode current, $A$ |
| $I_{Sh}$ | Shunt current, $A$ |
| $R_S$ | Series resistance, $\Omega$ |
| $R_{Sh}$ | Shunt resistance, $\Omega$ |
| $I_L$ | External load current, $A$ |
| $V_L$ | External load voltage, $V$ |
| $R_L$ | External load resistance, $\Omega$ |
| $V_T$ | PV cell thermal voltage (a curve fitting parameter), $V$ |

**Table 1.**
*Definitions of the operational parameters used in the PV-ECM.*

solar radiation level and PV cell temperature, the PV equivalent circuit model (PV-ECM) can be used to evaluate the power generation from the PV cell. **Table 1** presents the definitions of the important operational parameters related to the PV-ECM. The DC power generated (in $W$) by a PV cell is governed by Eq. (1):

$$P_{PV} = I_L\ V_L \tag{3}$$

$$I_L = I_{Light} - I_D - I_{Sh} \tag{4}$$

$$I_D = I_o\ \{exp\ [(V_L + I_L R_s)/V_T] - 1\} \tag{5}$$

$$I_{Sh} = \left[\frac{(V_L + I_L R_s)}{R_{Sh}}\right] \tag{6}$$

Eqs. (2)–(6) represent relationships between the various types of electrical currents for the purpose of estimating the load current. Their terminologies are presented in **Table 1** above (**Figure 3**).

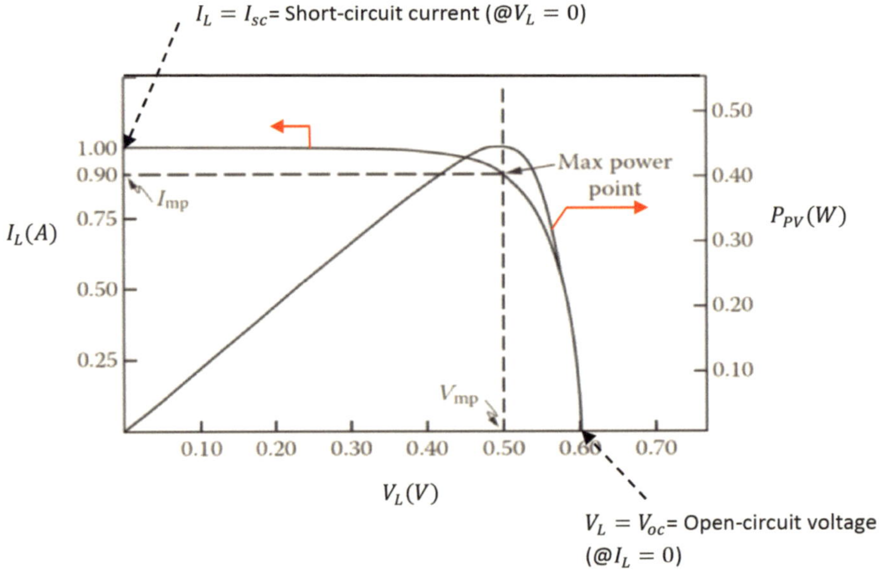


**Figure 3.**
*$I_L - V_L$ characteristic for a typical PV cell.*

$I_{Light}$, $I_D$ (or $I_o$), $R_s$, $R_{Sh}$, and $V_T$ may be function of PV cell temperature.

$$P_{mp} = I_{mp}\, V_{mp} \tag{7}$$

Traditionally, measurements of $I - V$ electrical characteristics for a PV cell are made at PV reference conditions of solar irradiance = $G_{ref}$ =*1000 W/m*$^2$ and PV cell operating temperature = $T_{C,ref} = 25^{\circ}C$. At short-circuit conditions (i.e., $I_L = I_{sc}$), $I_D \approx 0$, and when $R_{Sh} \to \infty$ (i.e., very large) $= > I_{Sh} = 0$. Using Eq. (2),

$$I_{Light} \approx I_{sc} \tag{8}$$

At open-circuit conditions (i.e., $I_L = 0$), so that

$$I_o = I_{Light}\; exp\,(-V_{oc}/V_T) \tag{9}$$

$$R_s = \frac{V_T\; ln\left(1 - \frac{I_{mp}}{I_{Light}}\right) - V_{mp} + V_{oc}}{I_{mp}} \tag{10}$$

The effect of cell operating temperature and solar radiation level on the *I-V* characteristic of a typical solar PV cell are shown in **Figures 4** and **5** [4], respectively.

To correct for the I-V characteristics based on varying the cell temperature or radiation levels, the following models can be used [1–5]:

$$V_T = V_{T,ref}\,\left(T_c/T_{c,ref}\right) \tag{11}$$

In Eq. (11), cell temperature must be in absolute temperature K. The light current and the reverse saturation current are given by

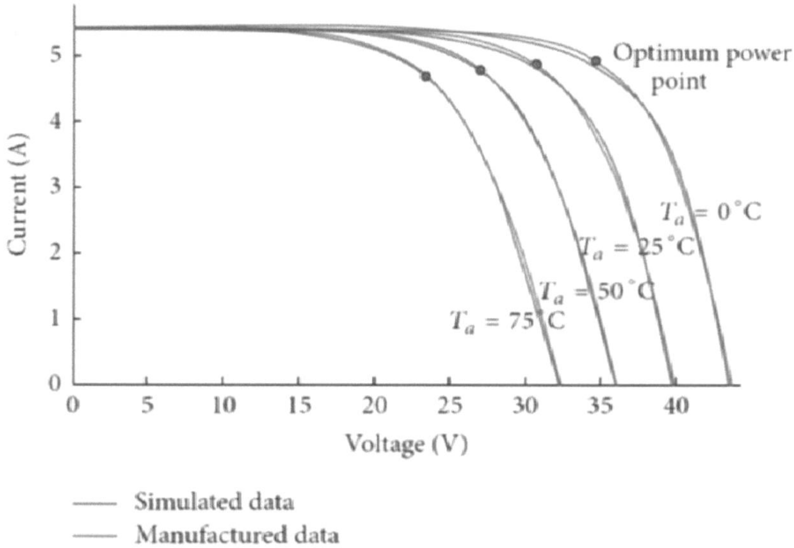


**Figure 4.**
*Typical I-V characteristics for a PV cell as function of cell operating temperature [4].*

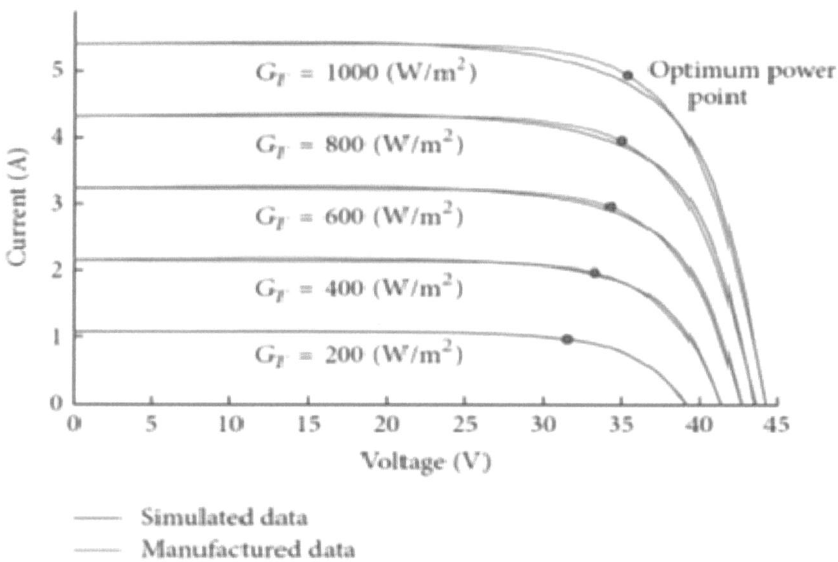


**Figure 5.**
*Typical I-V characteristics for a PV cell as function of absorbed solar radiation intensity [4].*

$$I_{Light} = \frac{G}{G_{ref}}\left[I_{Light,ref} + \mu_{I,sc}\left(T_c - T_{c,ref}\right)\right] \tag{12}$$

$$I_o = I_{o,ref}\left(\frac{T_c}{T_{c,ref}}\right)^3 exp\left[\frac{\epsilon N_s}{V_{T,ref}}\left(1 - \frac{T_{c,ref}}{T_c}\right)\right] \tag{13}$$

Where $\epsilon$ is the PV cell material bandgap energy (e.g., for silicon $\epsilon \approx 1.12\ eV$). $N_s$ is the total number of PV cells in series in a module times the number of modules in series.

$$R_s \approx R_{s,ref} \text{ (not function of } T_c \text{ in this model)} \tag{14}$$

$$\mu_{I,sc} = \frac{I_{sc}(T_2) - I_{sc}(T_1)}{T_2 - T_1} \tag{15}$$

Where $\mu_{I,sc}$ $(A/K)$ is the temperature coefficient of the short-circuit current. $T_2$ and $T_1$ are two temperatures centered around $T_{c,ref}$.

$$\mu_{V,oc} = \frac{V_{oc}(T_2) - V_{oc}(T_1)}{T_2 - T_1} \tag{16}$$

In Eq. (16), $\mu_{V,oc}$ $(V/K)$ is the temperature coefficient of the open-circuit voltage. The thermal voltage at reference conditions is given by

$$V_{T,ref} = \frac{\left(\mu_{V,oc}\, T_{c,ref} - V_{oc,ref} + \epsilon N_s\right)}{\left\{\left[\left(\mu_{I,sc}\, T_{c,ref}\right)/I_{Light,ref}\right] - 3\right\}} \tag{17}$$

$$\mu_{P,mp} = \eta_{mp,ref}\left(\frac{\mu_{V,oc}}{V_{mp}}\right) \tag{18}$$

In Eq. (18), $\mu_{P,mp}$ $(1/K)$ is the temperature coefficient of the power at maximum-power point. It is usually a negative value. The power conversion efficiency for the PV cell is at the maximum-power point given by

$$\eta_{mp} = \frac{P_{mp}}{A_c\, G} \tag{19}$$

$$\eta_{mp} = \eta_{mp,ref} + \mu_{P,mp}\left(T_c - T_{c,ref}\right) \tag{20}$$

In Eq. (19), $A_c$ (in $m^2$) is the solar PV collector (face) area receiving the solar radiation.

## 2.1 Estimation of $I_{mp}$ and $V_{mp}$—*Predictive models*

$I_{mp}$ and $V_{mp}$ are either given directly or can be estimated using graphical method or predictive models. For short-circuit conditions ($V_L = 0$), Eq. (2) can be approximated using [1–5],

$$I_L = I_{sc} - I_o\left[\exp\left(\frac{eV_L}{\kappa T_c}\right) - 1\right] \tag{21}$$

Where $e$ is the charge of an electron = $1.602 \times 10^{-19}$ coulombs (or $J/V$) and $\kappa$ is Boltzmann's constant = $1.381 \times 10^{-23}\ J/K$. For open-circuit conditions ($I_L = 0$ and $V_L = V_{oc}$), using Eq. (21), gives

$$0 = I_{sc} - I_o\left[\exp\left(\frac{eV_{oc}}{\kappa T_c}\right) - 1\right] \tag{22}$$

Rearranging Eq. (22) yields

$$V_{oc} = \frac{\kappa T_c}{e}\left[\ln\left(\frac{I_{sc}}{I_o}\right) + 1\right] \tag{23}$$

In order to determine the maximum power output $P_{mp}$, differentiate Eq. (1) with respect to $V_L$ and set the derivative equal to zero, that is,

$$\frac{dP_L}{dV_L} = I_L dV_L + V_L dI_L = 0 \tag{24}$$

$$\frac{dI_L}{dV_L} = -\frac{I_{mp}}{V_{mp}} \tag{25}$$

Also,

$$\exp\left(\frac{eV_{mp}}{\kappa T_c}\right)\left(1 + \frac{eV_{mp}}{\kappa T_c}\right) = 1 + \frac{I_{sc}}{I_o} \tag{26}$$

In addition, the values of $I_{mp}$ and $V_{mp}$ must satisfy

$$I_{mp} = I_{sc} - I_o\left[\exp\left(\frac{eV_{mp}}{\kappa T_c}\right) - 1\right] \tag{27}$$

Combining Eqs. (26) and (27), $I_{mp}$ and $P_{mp}$ can be determined to be

$$I_{mp} = \frac{\left(\frac{eV_{mp}}{\kappa T_c}\right)}{\left[1 + \left(\frac{eV_{mp}}{\kappa T_c}\right)\right]}(I_{sc} + I_o) \tag{28}$$

$$P_{mp} = \frac{\left(\frac{eV_{mp}^2}{\kappa T_c}\right)}{\left[1 + \left(\frac{eV_{mp}}{\kappa T_c}\right)\right]}(I_{sc} + I_o) \tag{29}$$

The use of Eqs. (28) and (29) will be explained in *Example 1*.

The power fill factor of a PV solar cell (or module), $PFF_{PV}$, is an interesting parameter, which can be used to assess the actual maximum power to the ideal (upper bound) power value, is given by

$$PFF_{PV} = \frac{I_{mp}\,V_{mp}}{I_{sc}\,V_{oc}} = \frac{P_{mp}}{I_{sc}\,V_{oc}} \tag{30}$$

In general, $I_{light} \approx I_{sc}$.

## 3. Configurations of solar PV cell arrangements

Solar PV cells, modules, and panels are typically connected electrically in series and/or parallel circuits to generate higher voltages, currents, and power outputs. However, complexities may arise because typically large quantities of power are

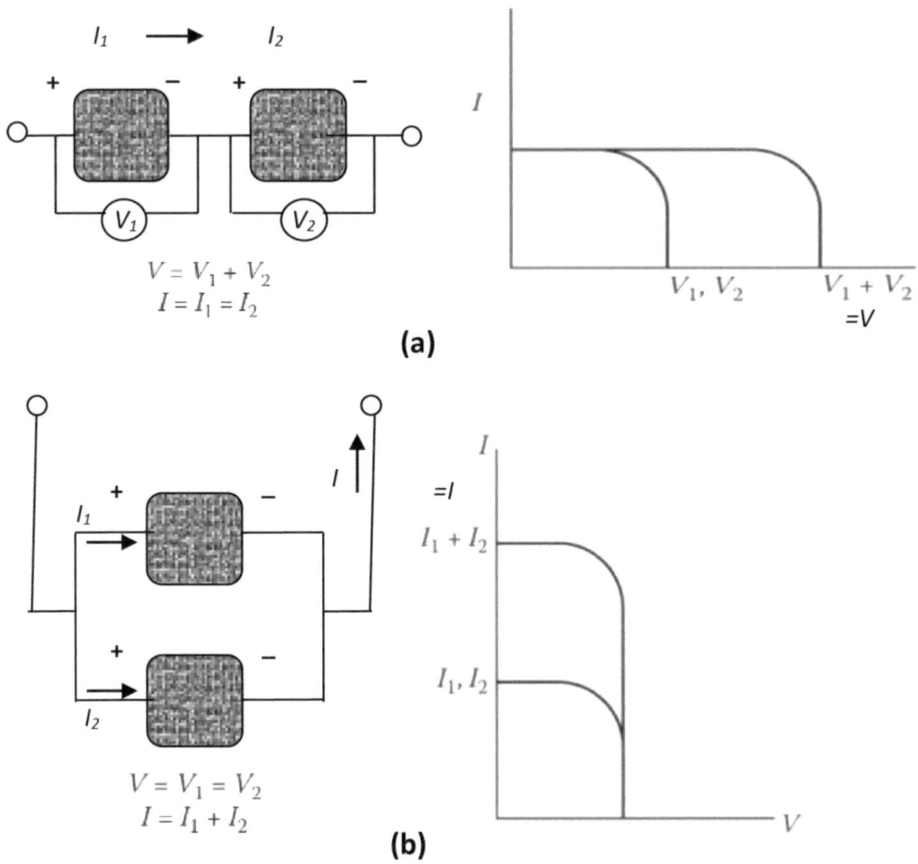


**Figure 6.**
*Electrical I-V Characteristics of two identical PV cells connected in: (a) series, and (b) parallel.*

required at specific voltages, and at a time and level of uniformity. Series and parallel configurations of PV cells follow the same rules as series and parallel DC circuits: (1) for identical cells arranged in series, the voltages add at constant current. (2) For identical cells arranged in parallel, the currents add at constant voltage. The electrical I-V characteristics of these arrangements are graphically shown in **Figure 6**.

## 4. Storage of PV generated electrical power: *Fundamental performance aspects*

Solar energy is an intermittent and time-dependent source. Solar PV systems are typically equipped with battery banks to store generated electrical power for use during times when their solar radiation is not available or at times when there is more demand for power generation using excess power stored. A battery is an electrochemical energy converter. It converts chemical energy into electrical power during the discharging cycle of the battery, that is, when battery is needed to supply power. It then converts electrical energy received from the PV system into chemical energy, in form of chemical reaction, during the charging cycle. Solar PV systems are used to generate power with varying amounts depending on the size of the PV system used. There are four types of PV systems, namely:

1. Direct-coupled PV system,
2. grid-connected PV system,
3. stand-alone PV system, and
4. hybrid PV system in which an auxiliary power source is integrated.

In general, the main components of a PV system may include:

1. PV convertor,
2. DC-AC power inverter,
3. battery bank (storage) and charge controller,
4. power conditioner, and
5. electric load.

A typical stand-alone PV power generation system integrated with a battery bank is shown in **Figure 7**.

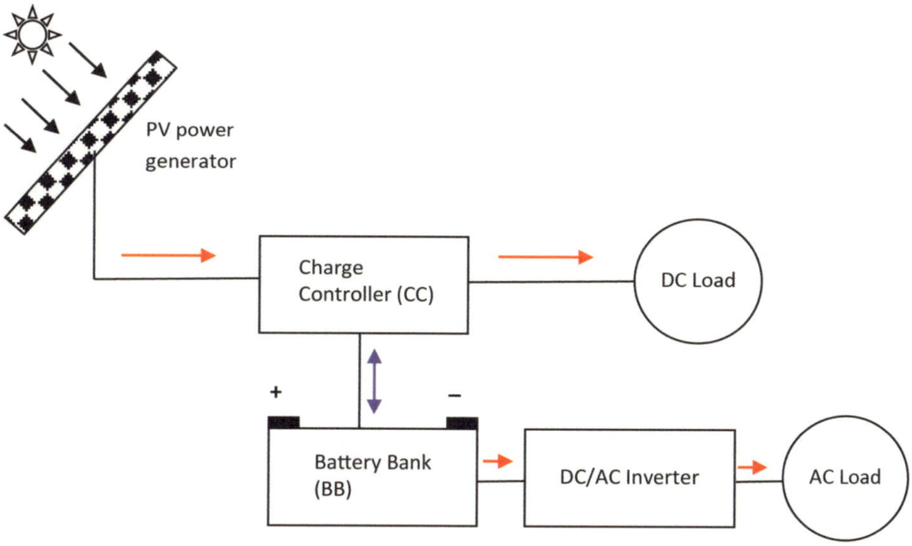


**Figure 7.**
*A stand-alone solar PV system with a battery storage system.*

| Battery operation parameter | Definition |
|---|---|
| $V_B$ | Battery terminal voltage, *V* |
| $C_B$ | Charge capacity, *Ah* *(amps-hour)* |
| *CC* | Cycle capacity (battery life indicator) |
| *DOD* | Depth of discharge, in %<br>DOD = 100% (battery is empty)<br>DOD = 0% (battery is full) |
| $E_B$ | Energy stored, *Wh or kWh* |

**Table 2.**
*Performance parameters used for battery systems.*

## 4.1 Battery performance characteristics

The basic parameters used in characterizing battery performance with their definitions are listed in **Table 2.**

$V_B$ and $C_B$ are used to specify the energy storage/retrieval potential of the battery system. The energy stored in the battery is estimated using:

$$E_B = V_B\,C_B \tag{31}$$

The battery cycle capacity refers to the number of charge/discharge cycles expected from the battery. The depth of discharge (DOD) specifies how much (in terms of %) of the stored energy at full capacity may be extracted without damage to the battery. Battery used in PV applications requires high cycle capacity as well as high DOD capability (i.e., deep cycle). It should be noted that automotive batteries are not suited for PV applications since they require low DOD capability. They typically require a large amount of power draw over a short period of time, called engine cranking time. PV systems with battery backup power bank require detailed economic feasibility as battery systems are typically expensive components. The total battery storage capacity required, $E_B$, for a given application is given by

$$E_{B,tot} = \frac{Energy\ storage\ requirement}{DOD} \tag{32}$$

The number of battery banks required can be estimated using

$$N_B = \frac{E_{B,tot}}{E_{B,one\ set}} \tag{33}$$

# 5. Illustrative numerical examples and case studies

## 5.1 Example 1

Draw the equivalent circuit for this PV module and estimate the power produced by the module under the given characteristics listed in **Table 3.**

| PV operation parameter | Value |
|---|---|
| $I_{Light}$ | 13.6 $A$ |
| $I_o$ | 0.008 $A$ |
| $V_T$ | 23.6 $V$ |
| $R_S$ | 0.9 $\Omega$ |
| $R_{Sh}$ | ∞ (very large) |

**Table 3.**
*Operating conditions for a given PV module.*

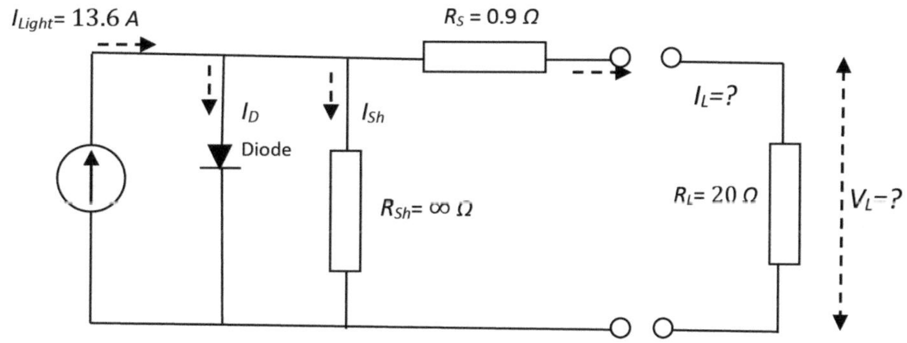


**Figure 8.**
*A schematic diagram showing the equivalent circuit of the given PV cell.*

### 5.1.1 Analysis

The equivalent circuit diagram for this case is shown in **Figure 8**.

The power generated by the given PV module can be estimated using Eq. (1) once $I_L$ and $V_L$ are determined. Substituting the known values in Eq. (21) gives

$$\mathrm{I_L} = 13.6 - 0.008\,\{\exp[(\mathrm{V_L} + 0.9\,\mathrm{I_L})/23.6] - 1\}$$

Using Ohm's law,

$$V_L = I_L\,R_L \tag{34}$$

Substituting back, gives

$$\mathrm{I_L} = 13.6 - 0.008\,\{\exp[(20\,\mathrm{I_L} + 0.9\,\mathrm{I_L})/23.6] - 1\}$$

Simplifying, yields

$$\mathrm{I_L} = 13.608 - 0.008\,\mathrm{e}^{0.886\,\mathrm{I_L}}$$

The value of $I_L$ can be estimated using trial and error procedure. The results are shown in **Table 4**.

Hence, $I_L = 7.5\,A$. Substituting back, gives $V_L = 150\ V$. Substituting back gives, $P_{PV\ module} = 1.125\ kW$.

| Guessed $I_L$ value = L.H.S. | R.H.S. result |
|---|---|
| 10.0 | −42.75 |
| 9.0 | −9.63 |
| 8.0 | 4.03 |
| 7.0 | 9.66 |
| 7.5 | 7.5 |

**Table 4.**
*Trial-and-error results.*

## 5.2 Example 2

### 5.2.1 Case A

Estimate the conversion efficiency of a 0.437 $m^2$ PV module at the maximum-power point if its maximum-power current and voltage are 2.50 $A$ and 16.5 $V$, respectively. The module operates at reference conditions of $G_{ref}$ =1000 $W/m^2$ and $T_{C,ref} = 25^{\circ}C$.

### 5.2.2 Analysis

The maximum power is estimated using Eq. (7)

$$P_{mp} = 2.5 \text{ x } 16.5 = 41.25 \text{ W}$$

Using Eq. (19) yields.

$$\eta_{mp} = \frac{41.25}{0.437\ x\ 1000} = 0.0944 \text{ (i.e.9.44\%)}$$

### 5.2.3 Case B

What would be the maximum power conversion efficiency of the PV module if the actual operating cell temperature of the module 67°C, if the temperature coefficient for the open-circuit voltage is $-0.0775\ V/K$.

### 5.2.4 Analysis

For this case, $\mu_{V,oc} = -0.0075\ V/K$ and $T_C = 67^{\circ}C + 273 = 340\ K$. Recall the results from the previous example, $\eta_{mp,ref} = 0.0944$ at $T_{C,ref} = 25^{\circ}C$+273 = 298 K, and $V_{mp,ref} = 16.5\ V$. Using Eq. (16), gives

$$\mu_{P,mp} = 0.0944\left(\frac{-0.0775}{16.5}\right) = -0.000442 \text{ (1/K)}$$

Using Eq. (20) yields.

$$\eta_{mp} = 0.0944 + -0.000442(340 - 298) = 0.075 \text{ (i.e.7.5\%)}$$

This means that there is a loss of approximately 2% due to operating the PV cell at a temperature higher than the reference temperature by $42^{\circ}C$.

### 5.3 Example 3

A 1.0 $m^2$ solar PV module made of silicon is operating at a cell temperature of $38^{\circ}C$ and receiving solar irradiance of 900 $W/m^2$. The short-circuit current and reverse saturation current were measured to be 200 $A$ and 1.8 x $10^{-8}$ $A$, respectively. Estimate:

i. The open-circuit voltage.

ii. The voltage at maximum-power point.

iii. The current at maximum power point.

iv. The maximum power generated by the PV module.

v. The maximum efficiency.

vi. The power fill factor.

#### *5.3.1 Analysis*

i. The open-circuit voltage for the PV module is estimated using Eq. (23),

$$V_{oc} = \frac{1.381 \text{ x } 10^{-23} \ (38 + 273)}{1.602 \text{ x } 10^{-19}} \left[ \ln\left(\frac{200}{1.8x10^{-8}}\right) + 1 \right] = 0.647 \ V$$

ii. The maximum power point voltage can be calculated using Eq. (26),

$$\exp\left(\frac{1.602 \text{ x } 10^{-19} \ V_{mp}}{1.381 \text{ x } 10^{-23} \ (38 + 273)}\right)\left(1 + \frac{e1.602 \text{ x } 10^{-19} \ V_{mp}}{1.381 \text{ x } 10^{-23} \ (38 + 273)}\right) = 1 + \frac{200}{1.8x10^{-8}}$$

Simplifying and solving for $V_{mp}$ gives

$$\exp(37.3 \ V_{mp})(1 + 37.3 \ V_{mp}) = 1.111x10^{10}$$

$$\mathrm{V_{mp}} = \frac{1.111\mathrm{x}10^{10} \ \mathrm{e}^{-37.3 \ \mathrm{V_{mp}}} - 1}{37.3}$$

The value of $V_{mp}$ can be estimated using trial-and-error method. The results are given in **Table 5**.

| The guessed value $V_{mp}$ = L.H.S. | R.H.S. |
|---|---|
| 1.0 | 188 |
| 0.5 | 2.30 |
| 0.60 | 0.030 |
| 0.55 | 0.34 |
| 0.54 | 0.506 |
| 0.53 | 0.747 |
| 0.536 | 0.592 |
| 0.537 | 0.569 |
| 0.538 | 0.547 |
| 0.539 | 0.526 |
| 0.5385 | 0.5365 |

**Table 5.**
*Trial-and-error results.*

Therefore, $V_{mp} \approx 0.5385\ V$.

iii. The PV maximum power point current can be calculated using Eq. (28),

$$I_{mp} = \frac{\left(\frac{1.602 \text{ x } 10^{-19}\ x\ 0.5385}{1.381 \text{ x } 10^{-23}\ (38+273)}\right)}{\left[1+\left(\frac{1.602 \text{ x } 10^{-19}\ x\ 0.5385}{1.381 \text{ x } 10^{-23}\ (38+273)}\right)\right]}\left(200 + 1.8x10^{-8}\right) = 190.5\ A$$

iv. The maximum output power can be estimated using Eq. (29),

$$P_{mp} = \frac{\left(\frac{1.602 \text{ x } 10^{-19}\ x\ 0.5385^2}{1.381 \text{ x } 10^{-23}\ (38+273)}\right)}{\left[1+\left(\frac{1.602 \text{ x } 10^{-19}\ x\ 0.5385}{1.381 \text{ x } 10^{-23}\ (38+273)}\right)\right]}\left(200 + 1.8x10^{-8}\right) = 102.6\ W$$

Or, using Eq. (7)

$$\mathrm{P_{mp}} = 190.5 \text{ x } 0.5385 = 102.6 \text{ W}$$

v. The PV conversion efficiency at maximum power point can be calculated using Eq. (19),

$$\eta_{mp} = \frac{102.6}{1.0\ x\ 900} = 0.114 \text{ (i.e.11.4\%)}$$

vi. The power fill factor can be estimated using Eq. (30),

$$PFF_{PV} = \frac{102.6}{200\ x\ 0.647} = 0.793 \text{ (i.e.79.3\%)}$$

### 5.4 Example 4

Design an arrangement of PV cells to provide an output of 12 $V$ and a power of 120 $W$. The maximum power current and voltage per cell are 5.13 A and 0.493 $V$, respectively.

#### *5.4.1 Analysis*

The maximum power output for a single cell is calculated using Eq. (7):

$$P_{mp} = 5.13\, x\ 0.493 = 2.529\ \text{W/cell}$$

The number of PV cells required to match the power output is the total power output required divided by the power output from a single cell. That is,

$$N_{power} = \frac{P_{total}}{P_{cell}} \tag{35}$$

$$N_{power} = \frac{120}{2.529} \approx 48\ \text{cells}$$

The number of PV cells required to match the specified voltage is the total voltage required divided by the voltage for a single cell. That is,

$$N_{voltage} = \frac{V_{total}}{V_{cell}} \tag{36}$$

$$N_{voltage} = \frac{12}{0.493} \approx 25\ \text{cells}$$

The proposed PV cell arrangement that matches these 2 requirements is shown in **Figure 9**.

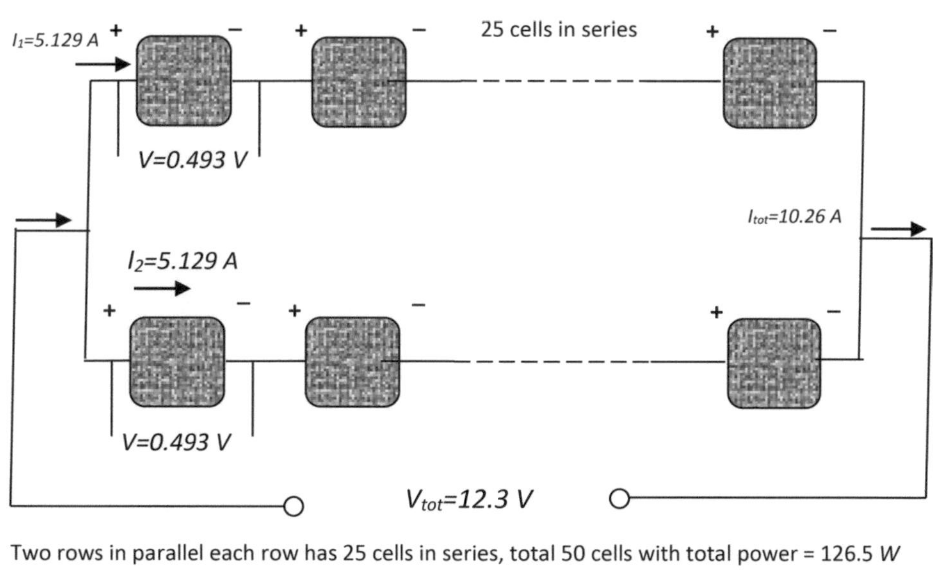


**Figure 9.**
*The proposed PV cells arrangement. Two rows in parallel each row has 25 cells in series, total 50 cells with total power = 126.5 W.*

### 5.5 Example 5

Consider a stand-alone PV system where it is needed to store and retrieve 10 kWh/day for 4 days with a depth of discharge of 50%. A single battery bank has a charge capacity of 1275 Ah. The voltage across the battery terminals is 12 V. Estimate how many battery sets would be required to meet the storage specifications. Show the configuration of the battery connection to the PV system.

#### *5.5.1 Analysis*

The total storage requirement = 10 kWh/day x 4 days = 40 *kWh.*
Using Eq. (32), the required total energy stored is estimated as.

$$E_{B,tot} = \frac{40}{50\%} = 80 \text{ kWh}$$

The energy stored in a single battery set is calculated using Eq. (31),

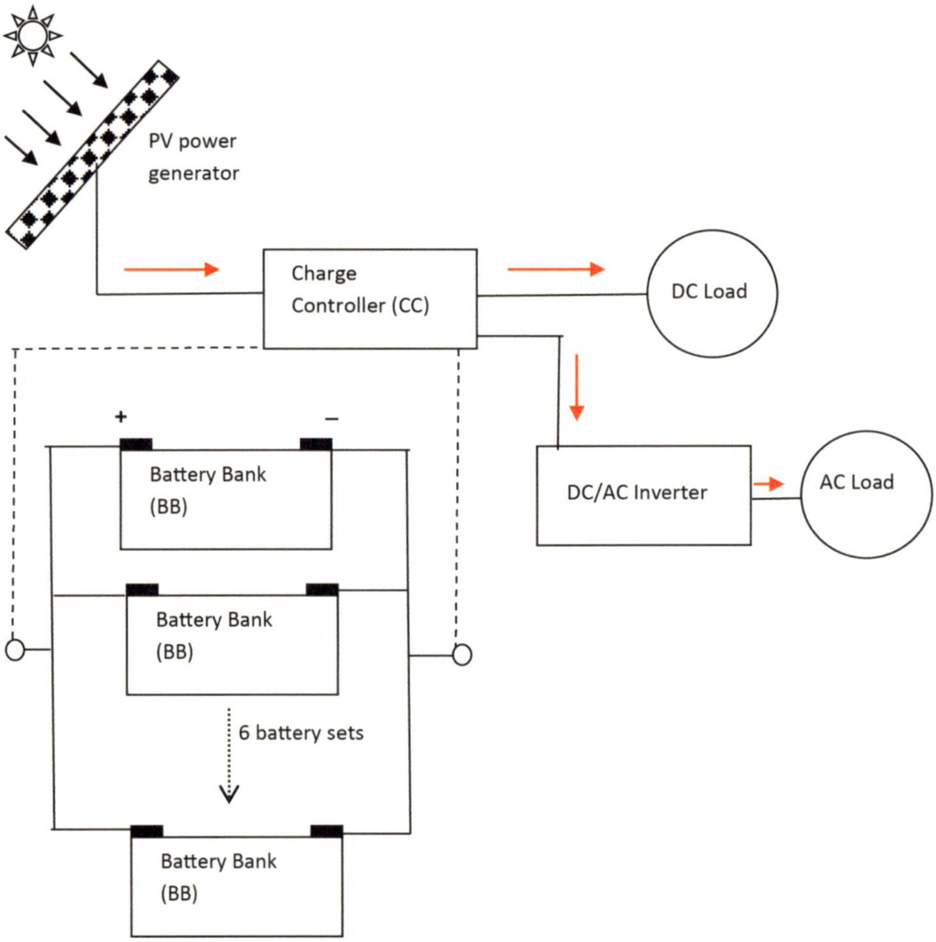


**Figure 10.**
*The proposed battery bank arrangement (in parallel configuration).*

$$E_{B,\text{one set}} = 12\ \text{V x } 1275\ \text{Ah} = 15.3\ \text{kWh}$$

The number of battery sets is estimated using Eq. (33),

$$N_B = \frac{80}{15.3} = 5.2 \approx 6\ \text{battery sets}$$

The results and the proposed PV system layout are depicted in **Figure 10**.

## 6. Conclusion

For many years, solar photovoltaic (PV) has proven and continued to be successful and promising source of renewable energy for power generation. In this chapter, fundamental aspects and modeling of power generation using solar PV systems are presented and discussed in detail. Also, illustrative numerical case studies based on fundamental energy conversion aspects and modeling are discussed.

## Author details

Basel I. Abed Ismail
Department of Mechanical and Mechatronics Engineering, Lakehead University, Thunder Bay, Ontario, Canada

*Address all correspondence to: baseliai@gmail.com

## References

[1] Goswami DY. Principles of Solar Energy. 3rd ed. Boca Raton, Florida, USA: Taylor and Francis Group, CRC Press; 2015

[2] Duffie JA, Beckman WA. Solar Engineering of Thermal Processes. 4th ed. Hoboken, New Jersey, USA: John Wiley & Sons Inc; 2013

[3] Patel MR. Wind and Solar Power Systems: Design, Analysis, and Operation. 2nd ed. Kings Point, New York, USA: Taylor and Francis Group, CRC Press; 2006

[4] Jakhrani AQ et al. An improved mathematical model for computing power output of solar photovoltaic modules. International Journal of Photoenergy. 2014;**2014**, Article ID: 346704:1-9

[5] Sumathi S, Kumar LA, Surekha P. Solar PV and Wind Energy Conversion Systems. Green Energy and Technology. Switzerland: Springer International Publishing; 2015

Chapter 3

# Energy Conservation Measures in Zero-Energy Building

*Tesfahun Meshesha*

## Abstract

Energy consumption of large public buildings has exceeded 33% of the total world energy consumption. Since it can help us lower the energy consumption of huge buildings to attain a zero-energy building (ZEB), energy conservation of large public buildings has attracted everyone's attention. The chapter is classified into five sections. The first two sections provide a succinct explanation of zero-energy building and a broad overview of energy conservation. The chapter then goes on to discuss ASHRAE Standard 90.1, which is a set of professional guidelines that the American Society of Heating, Refrigeration, and Air Conditioning Engineers regularly publishes to standardizes the demands placed on buildings to guarantee a low energy usage. Section four is all about different types of heat recovery systems that can reduce a building's energy usage. The last topic deals about what you need to do in your daily lives to use as little energy as possible. Energy-efficient buildings that produce enough renewable energy to meet their own annual energy consumption requirements is known as zero-energy building. Upon using different energy conservation practices discussed under this chapter along with the advancement of technologies, you can save huge amount of energy which will highly cut down your energy demand.

**Keywords:** Energy consumption, zero-Energy building, Energy conservation, ASHRAE 90.1, and heat recovery

## 1. Introduction

There are numerous social and economic variables and drivers that affect energy usage. The rapid population growth and expected considerable rise in domestic GDP, particularly in less developed nations, could lead to a significant rise in energy demand [1].

The current energy consumption of large public buildings in society has reached 1/3 of the total energy consumption [2], and buildings are central to the EU's energy efficiency policy, as they account for nearly 40% of final energy consumption [3]. The major areas of energy consumption in buildings are heating, ventilation, and air conditioning with 35% of the total building energy, lighting 11%, major appliances (water heating, refrigerators and freezers, and dryers) 18%, and the remaining 36% in miscellaneous areas including electronics [4]. In 2040, it is anticipated that the residential and commercial sectors will consume 20 percent more energy globally, with electricity serving as the primary energy source for the construction industry.

Electricity will provide for almost 90% of the growth [1]. Therefore, the energy conservation in large public buildings will be one of the important considerations for energy conservation and low-carbon development in countries around the world [2]. In all areas of the economy, there are numerous ways to lower energy use and boost efficiency. Achieving an efficient building which do have a zero-energy demand is the key in reducing the indicated energy consumption expected in the coming years. Buildings with zero-energy demand are known as zero-energy building (ZEB).

Zero-energy building produces enough renewable energy to meet its own annual energy consumption requirements, thereby reducing the use of non-renewable energy in the building sector [5]. ZEBs include renewable energy systems that can supply enough energy to cover residual needs while also reducing energy use through energy efficiency.

It is easier and less expensive to supply the building's energy needs with renewable sources of energy when energy consumption is reduced. ZEBs have the potential to completely change how buildings utilize energy, and more and more building owners seek to achieve this goal.

To achieve their business objectives, private commercial property owners are interested in creating ZEBs, and some have even built buildings with zero-energy demand. Even though, there is ambiguity and a lack of knowledge about what a ZEB is due to variation of definitions from region to region and from organization to organization, numerous state and local governments as well as federal agencies are beginning to move toward ZEB targets to meet regulatory obligations.

Moving toward ZEBs has a number of long-term importance, including as fewer negative effects on the environment, lower operating and maintenance costs, higher resilience to power outages and natural catastrophes, and increased energy security. Integrated design, energy-efficient retrofits, lower plug loads, and energy conservation initiatives are just a few ways to reduce building energy consumption during new building construction or remodeling.

To ensure a ZEB strategy, European Union has launched a net zero-energy building (NZEB) strategy. The Energy Performance of Building Directive (EPBD) states that Member States of European Union shall ensure that all new buildings are NZEBs by December 31, 2020 [6].

Technical measures for a renovated building in EU member states were divided into eight major intervention areas that are thought regarded as the NZEB's common denominator: Envelope, ventilation, lighting, control system, and renewable energy sources are the basic denominators [7]. Apart from eight focus areas of the technical measures taken by EU member states, there are also different energy conservation measures to which each individual person should be adapted on daily basis.

ZEBs should be implemented in emerging nations, notably those in Africa and Asia, where the energy consumption is anticipated to rise. Energy-saving companies (ESCO) have so far been seen to operate successfully in Germany, the United States, Hungary, China, and Brazil, but have had less success in some other nations [8]. Most ESCO projects in developing countries have been financed by bilateral and multilateral donors.

Building energy codes and standards play a significant role in achieving building energy efficiency by establishing minimum criteria for energy-efficient building design, construction, operation, and management. However, a study on building energy regulations found that 42% of the surveyed nations, the majority of which were developing nations, have no building energy standards [9].

In this chapter, some of the basic energy conservation measures and heat recovery mechanisms by which the building energy consumption can be reduced are going to be discussed. In addition, ASHRAE Standard 90.1, which is the energy standard that provides minimum requirements for energy-efficient designs for buildings except for low-rise residential buildings, is presented.

## 2. Energy conservation

These days, we hear the phrase "energy conservation" more and more frequently. Unfortunately, a lot of the places where we hear about it are in advertisements for goods or ways of living that might not actually be energy-efficient. We must start to comprehend the conservation theory that underlies them in order to discover what actual energy conservation techniques are.

Making limited resources last as long as possible will simply serve to exacerbate the problem until all energy sources are eventually depleted, which is the opposite of what energy conservation is meant to do. Reducing demand on a finite supply while allowing it to start rebuilding itself is the most effective method to the conservation strategy.

Finding new ways to access the Earth's supply of fossil fuels can also be considered as conservation in this scenario so that the regularly used oil sources are not fully depleted. As a result, such fields may replace themselves more frequently. When we talk about replenishing natural resources, we are talking about reducing the excessive demand on the supply over a long period of time so that nature may have sufficient time to recover.

Without a decrease in energy consumption, natural resources will eventually run out. Although some people do not see that as a problem because it will take many decades to happen and they expect that there will be an alternative by the time the natural resource is gone, depletion also results in the creation of a massively destructive waste product that then has an impact on other forms of life. The goals of energy conservation strategies include lowering demand, protecting and replenishing resources, developing and exploiting alternative energy sources, and correcting damage from prior energy operations.

Every building should incorporate energy-saving practices, starting with the conception of the building's structure and continuing through its design, construction, usage, maintenance, and demolition [10].

Regardless of how energy-efficient the system was when it was first installed, it is very important that everyone engaged in the building's design, operation, and maintenance recognizes this. Energy efficiency will decline in the absence of good operation.

## 3. ASHRAE standard 90.1

To help in the potentials put forward by the peoples of today's world toward reducing the energy consumption in a building, different standards followed by different nations have been developed. ASHRAE is a professional organization by American Society of Heating Refrigeration and Air Conditioning Engineers. They develop a standard for different systems in relation to heating, refrigeration, and air conditioning. AHSRAE 90.1 is one of those standards developed to standardize building's energy.

The minimal requirements for energy-efficient designs for buildings other than low-rise residential buildings are outlined in ASHRAE Standard 90.1, which is an energy standard for buildings (i.e., single-family homes, multi-family buildings less than four stories high, mobile homes, and modular homes). The standard was published by the American Society of Heating, Refrigeration, and Air Conditioning Engineers, maintained by ANSI, and jointly sponsored by the Illuminating Engineering Society (IES).

In 1975, ASHRAE 90, the first standard, was released. Since then, it has undergone numerous revisions. Based on the rapid advancements in energy technology and the rising cost of energy, the ASHRAE Board of Directors decided to adopt the continuous maintenance standard in 1999. This makes it possible for it to be updated more than once a year. In 2001, the standard adopted the new designation ASHRAE 90.1. Since then, it has been revised in 2004, 2007, 2010, 2013, 2016, 2019, and 2022 to take into account newer, more effective technology.

With the exception of low-rise residential buildings, the standard aims to "establish minimum requirements for energy-efficient design of buildings." It is a minimum requirement, although certain energy-saving initiatives, like "Leadership in Energy and Environmental Design, LEED," promote designs to use less energy than the minimal requirement.

In general, there are two means or paths for building designers to comply with ASHRAE Standard 90.1:

## 3.1 Prescriptive path

All components of the building meet the minimum standards specified by ASHRAE 90.1. ASHRAE 90.1 includes prescriptive requirements for the following:

### *3.1.1 Building envelope (section 5)*

Walls, doors, windows, roofs, and floors make up the building envelopes. The building's airtightness and the thermophysical properties of the materials used in its construction are crucial thermal factors in the design of the building envelope [11]. The thermal efficiency of the structural envelope has a direct impact on its energy utilization.

Standard design requirement for minimum wall insulation, minimum roof insulation, roof reflectance, and minimum glazing performance are all discussed in this section.

### *3.1.2 HVAC (section 6)*

Due to the constrained options available to designers for equipment, there is a simplified solution for single-zone structures with less than 25,000 $ft^2$ and only one or two levels. As long as the building has a single zone and just one unit, the unit must adhere to a few simple energy-saving standards.

There are several specifications for minimum equipment efficiencies for bigger buildings in terms of energy efficiency ratio, coefficient of performance, and integrated part-load value, or "EER," "COP," and "IPLV" [10].

The mandatory provisions discussed in this section are equipment efficiencies, verification, and labeling requirements [12].

i) Equipment efficiencies

For a standard condition, equipment efficiencies of those shown in Tables 6.8.1-1 through 6.8.1-16 shall have a minimum performance at the specified rating conditions when tested in accordance with the specified test procedure [12]. Where multiple rating conditions or performance requirements are provided, the equipment shall satisfy all stated requirements unless otherwise exempted by footnotes in the table. Equipment covered under the Federal Energy Policy Act of 1992 (EPACT) shall have no minimum efficiency requirements for operation at minimum capacity or other than standard rating conditions. Equipment used to provide service water-heating functions as part of a combination system shall satisfy all stated requirements for the appropriate space heating or cooling category.

For nonstandard Conditions, Equipment not designed for operation at AHRI Standard 550/590 test conditions of 44.0°F leaving and 54.0°F entering chilled-fluid temperatures, and with 85.0°F entering and 94.30°F leaving condenser-fluid temperatures, shall have maximum full-load kW/ton (FL) and part-load rating requirements adjusted using the following equations [12]:

$$FLadj = FL \,/\, Kadj$$

$$PLVadj = IPLV.IP \,/\, Kadj$$

$$Kadj = A \,/\, B$$

*where,*

- $FL = Full\ load \frac{kW}{ton} value\ from\ Table\ 6.8.1-3$
- $FLadj = \max.full\ load \frac{kW}{ton} rating, adjusted\ for\ nonstandard\ condition$
- $IPLV.IP = IPLV.IPvalue\ from\ Table\ 6.8.1-3$
- $PLVadj$ = maximum NPLV rating, adjusted for nonstandard conditions
- $A = 0.00000014592 \times (LIFT)^4 — 0.0000346496 \times (LIFT)^3 + 0.00314196 \times (LIFT)^2 — 0.147199 \times (LIFT) + 3.9307$
- $B = 0.0015 \times LvgEvap + 0.934$
- $LIFT = LvgCond — LvgEvap$
- $LvgCond = condenser\ leaving\ fluid\ temperature\ (°F)\ at\ FL$
- $LvgEvap = evaporator\ leaving\ temperature\ (°F) at\ FL$

The ***FLadj*** and ***PLVadj*** values are only applicable for centrifugal chillers meeting all of the following full-load design ranges:

- $36.0°F \le LvgEvap \le 60.0°F$
- $LvgCond \le 115.00°F$
- $20.0°F \le LIFT \le 80.0°F$

ii) Verification of Equipment Efficiencies

Equipment efficiency information supplied by manufacturers shall be verified by one of the following [12]:

a. Equipment covered under EPACT shall comply with U.S. Department of Energy certification requirements.

b. If a certification program exists for a covered product, and it includes provisions for verification and challenge of equipment efficiency ratings then the product shall be listed in the certification program.

c. If a certification program exists for a covered product, and it includes provisions for verification and challenge of equipment efficiency ratings, but the product is not listed in the existing certification program, the ratings shall be verified by an independent laboratory test report.

d. If no certification program exists for a covered product, the equipment efficiency ratings shall be supported by data furnished by the manufacturer.

e. Where components such as indoor or outdoor coils from different manufacturers are used, the system designer shall specify component efficiencies whose combined efficiency meets the minimum equipment efficiency requirements in Section 6.4.1.

f. Requirements for plate-type liquid-to-liquid heat exchangers are listed in Table 6.8.1-8.

iii) Labeling

a. Mechanical Equipment

Mechanical equipment that is not covered by the U.S. National Appliance Energy Conservation Act (NAECA) of 1987 shall carry a permanent label installed by the manufacturer stating that the equipment complies with the requirements of Standard 90.1 [12].

b. Packaged Terminal Air Conditioners

Nonstandard-size packaged terminal air conditioners and heat pumps with existing sleeves having an external wall opening of less than 16 in. high or less than 42 in. wide and having a cross-sectional area less than 670 sq.in. shall be factory labeled as follows: "Manufactured for nonstandard-size applications only: Not to be installed in new construction projects".

Under the perspective path, this section discuss the requirements for Economizers, Simultaneous Heating and Cooling Limitation, Air System Design and Control, Hydronic System Design and Control, Heat Rejection Equipment, Energy Recovery, Radiant Heating Systems, Hot Gas Bypass Limitation, and Refrigeration Systems. Generally to size systems and equipment, heating and cooling system design loads must be established in accordance with ANSI/ASHRAE/ACCA Standard 183.

### *3.1.3 Domestic hot water (section 7)*

Mandatory provisions of this section includes:

- **Calculations of Load:** Service water heating system design loads must be determined for the purpose of sizing the systems and equipment in accordance with manufacturers' published sizing recommendations or generally accepted engineering standards and handbooks that are acceptable to the adopting authority (e.g., ASHRAE Handbook—HVAC applications). All hot water storage tanks, hot water supply boilers, and pool heaters must adhere to the requirements indicated in this section [13].

- **Equipment Efficiency:** All hot water supply boilers, pool heaters, and hot water storage tanks shall comply with the requirements set forth in Table 7.8. When several conditions are specified, each one must be fulfilled [12].

- **Service Hot Water Piping Insulation:** All piping used in a recirculating system, including the inlet, branch, outlet, and piping that is heated externally, must be insulated to the standards specified in Table 6.8.3–1 of this section [12].

- **Service Water Heating System Controls:** There must be temperature controls that enable storage temperatures to be adjusted from 120°F or lower to a maximum temperature suitable for the intended application.

- **Pools:** Here are the specifications for the swimming pool's time switches, pool cover, and heaters [12].

**Heat Traps:** The requirements to set the level of heat trap are also presented [12].

Buildings with High-Capacity Service Water Heating Systems, Service Water Heating Equipment, and Space Heating and Service Water Heating are all covered under perspective path of this section.

### *3.1.4 Power (section 8)*

All building power distribution systems, transformer effectiveness, automatic receptacle controls, and energy monitoring are covered by this section. Mandatory provision of this section includes the following:

- **Voltage Drop:** The feeder conductors and branch circuits combined shall be sized for a maximum of 5% voltage drop total.

- **Electrical Energy Monitoring:** In new construction, measurement equipment must be installed to track the electrical energy consumption of HVAC systems, interior lighting, exterior lighting, and receptacle circuits. These systems must be monitored separately in buildings with tenants, both for the entire building and (apart from shared systems) for each individual tenant [13]. All loads listed in Section 8.4.3.1 must have their electrical energy use recorded at least every 15 minutes and reported at least hourly, daily, monthly, and yearly.

- **Low-Voltage Dry-Type Distribution Transformers:** Transformers that are not covered by the Energy Policy Act of 2005 have no performance criteria in this section and are noted as exceptions for convenience. Low-voltage dry-type transformers must adhere to the provisions of the Energy Policy Act of 2005.

*3.1.5 Lighting (section 9)*

Buildings in the United States utilize, on average, 35% of their total energy for lighting [10]. Savings opportunities are huge because of this. Although the Standard only permits a particular number of watts per square foot, W/ft2, the designer is given some latitude in the calculations. The permitted W/ft2 can be determined either by the kind of structure or by individual room. The standard permits trading between rooms and between HVAC and lighting so long as the net energy cost for the year does not go beyond the permitted limit.

Mandatory provisions of this section are as follows:

- **Lighting control:** Building lighting controls shall be installed to meet the provisions of Sections *9.4.1.1*, *9.4.1.2*, *9.4.1.3*, and *9.4.1.4*. Lighting controls are performed for interior lighting, parking garage lighting, exterior lighting, and special applications.

- **Exterior building lighting power:** The sum of the base site allowance and the individual allowances for areas that are intended to be illuminated and are allowed in Table 9.4.2–2 for the relevant lighting zone constitutes the overall exterior lighting power allowance for all exterior building applications. The outdoor lighting power allowance established in accordance with this section may not be exceeded by the installed exterior lighting power indicated in accordance with Section 9.1.3. Only the outdoor lighting applications indicated in Table 9.4.2–2's "Tradable Surfaces" section are subject to trade-offs. Unless otherwise indicated by the local authorities, Table 9.4.2–1 is used to calculate the lighting zone for the exterior of the building [12].

- **Functional testing:** To guarantee that control hardware and software are calibrated, adjusted, programmed, and in correct working condition in accordance with the construction papers and manufacturer's installation instructions, lighting control devices and control systems must be tested.

- **Dwelling Units:** Not less than 75% of the permanently installed lighting fixtures shall use lamps with an efficacy of at least 55 lm/W or have a total luminaire efficacy of at least 45 lm/W.

*3.1.6 Other equipment (section 10)*

This section is basically used to specify the requirements for electric motors, potable water booster pumps, elevators, and escalators. Thus, the mandatory provisions of this section are as follows:

- **Electric motors:** Electric motors must meet the standards outlined in Table 10.8–1 for NEMA Design A motors, NEMA Design B motors, and IEC Design N motors, and Table 10.8–2 for NEMA Design C motors and IEC Design H motors, whether they are produced separately or as a part of another piece of equipment [12]. General-purpose small electric motors must have a minimum average full-load efficiency of not less than that specified in Table 10.8–3 for polyphase small electric motors and Table 10.8–4 for small electric motors with capacitor-start capacitor runs and capacitor-start induction runs. These motors must have a rated motor power of at least 0.2 hp. and less than or equal to 3 hp.

- **Service Water Pressure Booster Systems:** Service water pressure booster systems are designed such that the following shall apply [12]:

    a. To change pump speed and/or start and stop pumps, one or more pressure sensors must be employed. The sensors must either be placed close to the crucial fixtures that determine the necessary pressure or circuitry that modifies the set point to mimic the action of remote sensors must be used.

    b. Except for safety devices, no equipment may be installed with the intention of lowering the pressure of all water delivered by any booster system pump or booster system.

    c. When there is no service water flow, no booster system pumps may work.

- **Elevators:** Elevator systems must essentially state what is needed for the lighting, ventilation power limits, standby mode, and design documentation.

- **Escalators and moving walks:** When not carrying people, escalators and moving walkways must automatically slow to the minimum allowable speed in line with ASME A17.1/CSA B44 or the relevant local code.

- **Whole building energy monitoring:** At the construction site, instruments must be installed to measure how much energy each new structure uses.

### 3.2 Performance path

The two performance-based strategies in Standard 90.1 are the energy cost budget (ECB) technique, which is covered in Section 11, and the performance rating method (PRM), which is better known for its location in the Standard, Appendix G [12]. These techniques give more flexibility by allowing a designer to "trade off" compliance by forgoing some prescriptive criteria if the harm can be offset by exceeding other prescriptive requirements. This is shown by comparing a proposed building design to a reference building design, often known as a baseline, using computer simulation. The features of the base building design are the primary distinctions between the ECB and PRM approaches in Standard 90.1.

#### *3.2.1 Energy cost budget (ECB)*

The building energy cost budget method is an alternative to the prescriptive provisions of this standard. It may be employed for evaluating the compliance of all proposed designs except designs with no mechanical system [12].

- Most of the building components being adjusted to "just meet" the current prescriptive requirements, and the baseline is virtually duplicated in the proposed design of the building.
- When the yearly energy cost of the proposed design is equal to or less than the annual energy cost of the baseline building design, the structure is judged to be in compliance with the ECB. A dependent baseline is the term used to describe this strategy.

### *3.2.2 Appendix G*

- Many of the baseline design features in Appendix G which use more independent baseline are based on standard practice. Therefore, credit is given for surpassing both the code's prescriptive criteria and standards that are not subject to its regulation.
- The efficiency standards in Appendix G are established at values that are not meant to be modified with each new addition of the code using a stable baseline approach. Instead, the proposed building's energy performance needs to be higher than the industry standard by a margin appropriate for the code year under review.

## 3.3 ASHRAE Standard 90.1-2022

Many states mandate that newly constructed or renovated buildings comply with ASHRAE 90.1. The majority of states impose uniform or comparable criteria on all commercial structures. Others use the same or comparable principles for all governmental structures. Some jurisdictions use different energy-saving requirements for all commercial buildings, while others combine the ASHRAE 90.1 standard for all public buildings with different energy-saving requirements for private structures. For their governmental and commercial structures, a few states do not enforce any energy-saving standards.

The Building Codes Assistance Project keeps track of the adoption of energy codes at the moment. In September 2020, there will be 7 states with building codes that meet or exceed ASHRAE Standard 90.1–2016, 14 states with codes that meet or exceed 90.1–2013, 8 states with codes that meet or exceed 90.1–2010, 9 states with codes that meet or exceed 90.1–2007, and 8 states with either no statewide code or a code that predates 90.1–2004 [14].

The structure and specifications of the California Energy Code (CCR Title 24 Part 6) are remarkably similar [15]. The USGBC also makes reference to ASHRAE 90.1 as an industry standard in the LEED building certification process. It is widely utilized during energy retrofit projects or any project that makes use of building performance simulation as a baseline for comparison.

Since its start in the year 1975, ASHRAE Standard 90.1 has been modified over year, to incorporate more energy-saving practices. Buildings meeting the requirements will successfully save high amount of energy. New commercial buildings meeting the requirements of Standard 90.1–2019 were analyzed comparative to Standard 90.1–2016 and exhibit national savings of approximately 4.7% site energy savings, 4.3% source energy savings, 4.3% energy cost savings, and 4.2% carbon emissions [16]. The 2022 edition of ASHRAE Standard 90.1 being the latest, additional addenda are included to the 2019 edition.

ANSI/ASHRAE/IES 90.1–2022 helps meet the need by offering minimum energy-efficient requirements for the design, construction, operation, and maintenance of buildings and sites and their utilization of onsite renewable energy resources. It applies to new buildings and new portions of buildings and their systems, new systems and equipment in existing buildings, new systems and equipment that are part of a site, and new equipment or building systems that are part of industrial or manufacturing processes. The standard also provides criteria for determining compliance with these specifications.

Single-family homes, multi-family buildings with three stories or less above grade, manufactured homes (mobile homes), prefabricated homes (modular homes), and structures that do not use fossil fuels or electricity are exempt from the requirements of ANSI/ASHRAE/IES 90.1–2022. These buildings' energy efficiency requirements are covered in another document, ANSI/ASHRAE/IES 90.2–2018: Energy-Efficient Design of Low-Rise Residential Buildings [17], which can also be also found on the ANSI Webstore. Refer to "ANSI/ASHRAE 90.2-2018: Energy-Efficient Design of Low-Rise Residential Buildings" for more information about this standard.

The ANSI/ASHRAE/IES Standard 90.1–2022 Energy Standard for Sites and Buildings except Low-Rise Residential Buildings (I-P Edition) can be obtained at https://www.techstreet.com/ashrae/products/preview/2522082.

The 2022 version of ASHRAE Standard 90.1 includes more than 80 addenda to the 2019 version, along with a number of energy-saving improvements. Notable changes include the following:

### *3.3.1 General*

- The scope of the standard has been expanded to include sites as well as buildings, enabling regulation of energy use associated with the building but not in the building itself, such as exterior or parking lot lighting not connected to the building electric service. This also allows for equipment, such as photovoltaic (PV) equipment, located on site but not within the building.

- A new energy credits requirement (new Section 11) has been added that enables approximately 4–5% cost-effective energy savings through 33 different energy-saving measures. The number of required credits varies by building type and climate zone.

- A minimum prescriptive requirement for onsite renewable energy has been added. This requirement includes exceptions for small buildings, buildings with limited roof space, and other situations where PV installations would be problematic.

- It should be noted that, due to the addition of a new section and several appendices, the number or letter designations of several well-known sections and appendices have changed from those used in the 2019 and previous editions of the standard.

### *3.3.2 Building envelope*

- A requirement was added to perform whole building air leakage testing and measurement on buildings less than 25,000ft$^2$.

- Requirements were added that address the impacts of thermal bridges in building envelopes, with a new Informative Appendix K providing supplemental information on application.

- A solar reflectance requirement for walls was added for Climate Zone 0. This is similar to the requirements for high-albedo roofs.

- Specific provisions were added to distinguish roof replacements from other types of alterations.

- A new reference was added for steel-framed walls to allow use of ANSI/AISI S250 for U-factor determination.

- A definition was added for insulated metal panels (IMPs).

- Normative Appendix A was reformatted to clarify the requirements for thermal performance calculations.

*3.3.3 Lighting and power*

- Reorganized Section 9, "Lighting," to be more consistent with the structure of other main sections of the standard.

- Updated installed interior lighting power allowances and minimum control requirements; added a power exception for the germicidal function in luminaires and sources; and removed exceptions for casinos and parking garage daylight transition zone lighting.

- Modified a number of lighting requirements to reflect greater use of higher efficiency LED products and revised lighting practices.

- Added requirements for indoor horticultural lighting in greenhouses and indoor grow buildings based on a new metric, photosynthetic photon efficacy (PPE), developed in ANSI/ASABE S640.

- Provided an additional interior lighting power allowance for video conferencing. Power allowances and controls have been moved to a table for ease of reference.

*3.3.4 Mechanical*

- Introduced an optional Mechanical System Performance Path that allows HVAC system efficiency trade-offs based on a new metric, total system performance ratio (TSPR).

- Required condensing boilers for new construction in order to achieve 90% or greater efficiency for large boilers (1 to 10 million Btuh). The thermal efficiency requirements for high-capacity gas-fired service water heating equipment were also increased.

- Established a minimum enthalpy recovery ratio for energy recovery systems and specified operational requirements to ensure proper economized performance.

- Revised demand control ventilation requirements to be based on climate zone and Standard 62.1 airflow requirements.

- Modified the minimum efficiency requirements for air source heat pumps and introduced a new metric, $COP_{HR}$, for units that perform heat recovery during chiller operation.
- Added the minimum energy efficiency requirements (and new CFEI metric) for large-diameter ceiling fans from 10 CFR 430.

### *3.3.5 Performance rating method (appendix G)*

- New requirements were added to limit trade-offs between the building envelope and other building systems.
- A new Informative Appendix I was added that provides information on how alternative performance metrics other than cost could be used with the performance rating method. These alternative metrics include site energy, source energy, and carbon emissions, and would be useful to assess building performance against carbon emission goals or for similar types of comparison.
- A relaxation in stringency was added when using Normative Appendix G for retrofit projects consisting of substantial alterations.

### *3.3.6 Both performance-based compliance paths (section 12: Energy cost budget method and appendix G)*

- Numerous changes were included to improve clarity and coordinate with revisions to other sections of the standard.

### *3.3.7 New normative appendix J*

- Contains performance curves that represent minimally compliant chiller performance for the budget and baseline building design and for proposed building designs when specific equipment performance is not known.

## 4. Heat recovery

The majority of industrialized nations account for one-third of their entire energy usage in HVAC. Additionally, in hot and humid climates, cooling and dehumidifying fresh ventilation air accounts for 20 to 40% of the overall energy load for HVAC [18]. However, where complete fresh air ventilation is required, that proportion may be higher. This indicates that more energy is required to satisfy the occupants need for fresh air.

Due to rising energy costs associated with treating fresh air, heat recovery is becoming increasingly necessary. By recovering the waste heat, heat recovery systems can reduce the amount of energy needed to heat, cool, and ventilate buildings. In order to save energy, stand-alone or combined heat recovery systems can be added to residential or commercial structures. Reduced energy use can make a significant contribution to a sustainable society by lowering greenhouse gas emissions.

Buildings often have a core unit, fresh and exhaust air channels, and blower fans as part of their heat recovery systems. Depending on the temperature, the season, and the needs of the building, exhaust air from buildings is employed as a heat source or a heat sink. Heat recovery systems have considerably increased a building's energy efficiency by recovering typically between 60 and 95 percent of the heat in exhaust air [19].

Heat recovery ventilation (HRV), also known as mechanical ventilation heat recovery (MVHR), is an energy recovery ventilation system that uses two air sources with different temperatures. Buildings can use less energy for heating and cooling by recovering heat from existing energy sources. Before the fresh air enters the room, the air cooler of the air conditioning unit performs heat and moisture treatment. Due to this, fresh air fed into the air conditioning system is preheated (precooled) and the fresh air enthalpy is increased (decreased). This is done by extracting the remaining heat from the exhaust gas [18].

## 4.1 Types of heat recovery systems

### *4.1.1 Rotary thermal wheels*

Mechanical heat recovery is accomplished *via* rotary thermal wheels. By passing through each fluid alternately, a revolving porous metallic wheel transmits heat energy from one air stream to another. When the device is in operation, the wheel matrix acts as a thermal storage mass, briefly storing the heat from the air before it is transferred to the cooler air stream [18]. Heat wheels and enthalpy wheels are two different categories of rotary thermal wheels. Heat and enthalpy wheels have geometrical similarities, but there are also distinctions that have an impact on how each design works. The moisture in the air stream with the highest relative humidity in a system containing a desiccant wheel is transferred to the opposing air stream after passing through the wheel. Both incoming air to exhaust air and exhaust air to incoming air can be used in this. The air can then be immediately consumed or further cooled using the supplied air. This technique enables us to reuse a lot of energy [20].

### *4.1.2 Fixed plate heat exchangers*

The most popular kind of heat exchanger, which has been around for 40 years, is a fixed plate unit. A little gap separates each stack of thin metal plates. These areas are traversed by two separate air streams that are next to one another. Heat transmission occurs as temperature shifts occur from one air stream to another across the plate. These devices have demonstrated sensible heat efficiency rates of 90% while transferring heat from one air stream to another [21]. The high heat transfer coefficients of the materials utilized, operational pressure, and temperature range are responsible for the high levels of efficiency [19].

### *4.1.3 Heat pipes*

One way heat transferred from one air stream to another is by the employment of another heat recovery device known as heat pipes [19]. Heat is transmitted along the heat pipe utilizing an evaporator and condenser inside a wicked, sealed tube that contains a fluid that is continually changing its phase. This heat is then transferred to the heat pipe's two extreme ends. The liquid in the pipes turns into a gas and receives thermal energy from the heated air stream in the evaporator portion. The gas in the condenser part condenses back into a fluid, releasing thermal energy into the cooler air stream, which causes the temperature to rise. Depending on the configuration, the fluid or gas moves from one side of the heat pipe to the other using wick forces, pressure, or gravity.

#### *4.1.4 Phase change materials*

A building structure with phase change materials (PCMs) can store sensible and latent heat at a higher storage capacity than buildings with conventional building materials. Due to their capacity to retain heat and shift heating and cooling demands from traditional peak hours to off-peak times, PCMs have received substantial research. Long understood and researched is the idea of a building's thermal mass, or how the building's physical structure may absorb heat to assist cool the air. The thermal storage capacity of PCMs is 12 times more than that of conventional building materials across the same temperature range, according to a study comparing PCMs to traditional building materials [21]. There is a decrease in pressure across PCMs.

#### *4.1.5 Run-around systems*

Run-around systems are hybrid heat recovery systems that combine the features of existing heat recovery technologies to create a single unit that can collect heat from one air stream and transmit it to another a long way away. In the typical run-around heat recovery scenario, two fixed plate heat exchangers are placed in two different air streams and connected by a closed loop with a fluid that is continuously pumped between the two heat exchangers. Heat recovery is provided by the fluid's continuous heating and cooling as it circulates around the loop. Pumps must be able to move between the two heat exchangers due to the fluid's continual flow through the loop. Although there is an increased energy demand, moving fluid with pumps instead of moving air with fans uses less energy [22].

## 5. Energy conservation practices

In addition to designing and constructing buildings according to ASHRAE Standard 90.1 so that they comply with the requirements set by the standard to minimize energy consumption, you the peoples getting service from the buildings have to contribute your share to energy conservation to realize zero-energy building. The following are some of the practical energy conservation techniques that can help you to reduce your overall carbon footprint and save money in the long run.

### 5.1 Use Energy-efficient lighting

By replacing typical incandescent lights with long-lasting and energy-efficient LED lamps, energy consumption can be significantly reduced. Compared to incandescent bulbs, LEDs use up to 90% less energy [23]. Along with LED retrofits, smart lighting systems that adjust lighting based on variables like occupancy, time of day, and natural lighting conditions can help further reduce rising electricity costs.

### 5.2 Reduce the room's temperature

Energy can be saved significantly even with a little drop in room temperature, say, just one or two degrees. The amount of energy needed to keep a room at a given temperature depends on how much the inside and external temperatures differ. Purchasing a programmable thermostat would be a wiser and more pleasant course of action [24]. Rooms with low temperatures can be further achieved by using a cool roof. Cool roofs are built of highly reflective materials to reflect more light and absorb less heat from the sun, keeping homes cooler in hot weather [25].

### 5.3 Reducing air leakage

Air leaks that could be costing you additional money will be fixed by proper insulation [26]. If your home is not properly insulated during the winter, you could be losing a lot of heat. You can either conduct the leak repairs yourself or hire an energy professional to help you.

### 5.4 Use maximum daylight

Use natural light as much as you can by turning off your lights during the day. This will reduce the strain on the neighborhood power system and end up saving you a significant sum of money over time [23].

### 5.5 Perform Energy audit

A home energy audit is nothing more than a procedure that enables you to pin-point the places in your house where energy is being lost and the remedial actions you can take [27]. If you follow the advice provided by energy experts, your monthly electricity cost may decrease.

### 5.6 Use Energy-efficient appliances

Choose electrical equipment with an Energy Star rating above others when making a purchase decision. Appliances with an Energy Star rating use less energy and cost less to operate [27]. You will ultimately save money by investing in them, despite the fact that they initially cost more.

### 5.7 Drive less, walk more, and carpooling

Driving less and walking more is another energy-saving strategy. As walking is a fantastic form of exercise, this will not only lessen your carbon footprint but also keeps you healthy [25]. If you drive to work and some of your coworkers live nearby, consider carpooling. This will not only decrease your monthly fuel costs but also improve your social participation.

### 5.8 Turn off idle appliances

The majority of plug loads in commercial buildings are caused by computers, monitors, networking equipment, and office electronics. Plug loads make up about 25% of the overall energy consumption in office buildings that bare the minimum requirements of the construction code, and they can rise to 50% [28]. Although there is no a one optimum conservation technique to lower plug loads, the best course of action is to switch the equipment off completely after work hours.

### 5.9 Plant shady landscaping

When it is hot and sunny outdoors, shady landscaping will shield your house from extreme heat and chilly winds in the winter [27]. When you compute the amount of energy saved at the end of the year, it will eventually result in significant savings by keeping your home cool during the summer. Planting trees on the southern and western sides of your home will help keep it cooler since they will block the sun's rays from directly hitting it. In the winter, when the trees lose their leaves, more sunshine will enter your home [29].

### 5.10 Install Energy-efficient windows

We have some older windows in our homes that are not energy-efficient. Single-panel windows are significantly inferior to double-panel windows and other vinyl frames. Your power bills might be reduced by selecting the right blinds [25].

### 5.11 Bicycles as your best friend

Indeed, bicycles could be a big assistance to you in the effort to conserve energy. Bicycles are literally your best buddy because they are powered entirely by human effort and require no other kind of energy. The gain is simply doubled if you are health conscious [30]. Simply by choosing to commute between locations on bicycles, you are not only helping to protect the environment but also working toward leading a better lifestyle. Overall, it is advantageous for both you and energy conservation [25]. With all of this in mind, many nations have created designated bicycle lanes to encourage the usage of bicycles and to promote the safety of those riding them. The Netherlands was the first country to create a bicycle lane in Amsterdam.

### 5.12 Buying a programmable thermostat

Let us be honest, you have all experienced moments in your lives when you were aware that you did not require the heater or the air conditioner, but you chose to disregard that inner voice out of laziness. Life simply becomes a little bit more productive with a programmable thermostat. Even when you are away, a programmed thermostat can turn itself on and off as needed without bothering you. In addition to being cost-effective, it will also enable you to contribute to energy conservation [25].

### 5.13 Use motion detectors

In addition to lowering your energy bill budgets, installing motion detectors could greatly assist you in achieving the goal of energy conservation [25]. It could be quite beneficial to install motion detectors, especially for outdoor lighting. The result is that you no longer have to turn the key in complete darkness because you do not have to bother about turning them off when you leave or even worrying about turning them on when you arrive home. Your life would be a little bit more convenient with motion detectors, and they are effective at minimizing energy waste.

### 5.14 Closing doors

There is no way to be more literal than this. One of the best strategies to save energy is to close the doors right after you. This applies to the doors of the refrigerator and the rooms where the air conditioner is on [25]. In addition to saving the equipment, this lowers your electricity costs. More importantly than anything else, it is your small contribution to energy conservation.

## 5.15 Increase roof insulation

A commercial structure loses 20% of its heat through the roof [31]. These losses can be reduced by increasing the resistance capacity of the roof materials, and the cost of installing more insulation to the roof can be repaid in as little as 1 to 4 years. Building managers should not solely depend on prior R-value numbers to determine whether or not to add extra insulation, since the insulation can lose effectiveness over time as it becomes more compact. According to the Environmental Protection Agency, installing insulation in attics, crawl spaces, and basement rim joists can help homeowners save 15% on heating and cooling costs (or 11% of overall energy costs) [32].

## 5.16 Other Energy conservation techniques

While there are practical methods such as insulation, changing light sources, using alternate fuels, and carpooling rather than walking, there are also a few energy conservation techniques which will be surprising to us. These include:

***Education:*** Education is probably the most powerful of all energy conservation techniques that can be used. Tell everyone about your commitment to energy savings. Use your company newsletter and company or building announcements to keep tenants informed about your energy-saving goals and how they can both help and benefit [33]. Education is not just about teaching people the importance of conservation, but it is also about showing the alternative choices that can be used in construction, manufacturing, and other processes so that energy can highly be conserved [25].

***Zero-energy balance:*** Zero-energy balance encompasses more than just energy-saving strategies for environmentally friendly building. It is a process of reassessing and redesigning industrial and commercial activities, so they can gather and store energy as well as take it and put it back onto the grid to lessen the stress of brown outs [29].

***Alternative power:*** Alternative power and fuel sources are becoming more prevalent in a variety of life processes. Since almost all transition models call for the replacement of the current processes with more energy-efficient ones, the use of alternative power is one of the most important energy conservation strategies. In United Kingdom, solar energy represents around 1.2% of all electricity used currently [29]. This figure can be highly improved to harness a high amount of solar energy potential.

***Cap and trade agreements:*** Cap and trade agreements are a tool used to control and reduce pollution and consumption in the manufacturing sector. The companies are permitted to emit a specific amount of emissions, which they may bid to increase [25]. The extension bid is then applied to projects as compensation. Even while it might not seem like it, this is fundamentally about energy conservation.

***Reduced demand:*** To lessen the total demand on the world's energy supplies, several efforts are in place. This can involve anything from modifying the type of insulation required in new construction to educational initiatives [25].

***Research and development:*** Our ability to make changes to reduce consumption and find renewable sources of the energy needed for modern living depends on the support of ongoing research and development programs in the field of energy conservation [25]. Given its potential to help resolve the global energy issue, it ought to be considered one of the most valuable energy conservation strategies.

***Switching to solar energy:*** Solar energy is one of the most effective sources of energy. Available in huge amounts, especially in tropical regions, solar energy is a sustainable source of energy [25]. At any given moment, the Earth receives roughly

173 trillion kW of energy from the sun, which is 10,000 times more than the entire world population uses [34]. Only a tiny fraction of this potential is currently being used, but as technology develops and spending increases, much more may be tapped into [29]. It is true that there are some restrictions on using solar energy, such as the high initial cost. Furthermore, it can only be used on days with clear skies. However, converting to solar energy will significantly conserve our finite energy resources in addition to protecting the environment.

## 6. Conclusions

This chapter has covered a variety of energy conservation techniques used to create a zero-energy building. Complying with those energy-saving practices, considerable amount of energy can be saved over a period of time. ASHRAE Standard 90.1 provides the energy standards for a building except for low-rise residential buildings. Since residential building takes a higher share of energy consumption, while preparing for the construction of new buildings and remodeling, building designers must meet this standard to save as much energy as possible. By utilizing the fundamental types of heat recovery techniques covered in this chapter, energy-efficient buildings can be achieved as heat recovery is one of the primary strategies for reducing the energy demand of a building.

The behavior of building occupants has a big impact on energy demand, so giving suggestions for improving occupant behavior is quite helpful for energy saving. Every regular occurrence that transforms an appliance's inactive state into an active one aids in creating user energy consumption profiles based on their activities. The amount of energy wasted can be reduced by choosing an efficient timetable for appliance operation. In order to assist energy saving, 16 strategies that we should include into our daily activities have been discussed.

Zero-energy building can be realized, upon the implementation of different energy conservation measures discussed in this chapter.

## Acknowledgements

I am thankful for the almighty God.

## Conflict of interest

The authors declare no conflict of interest.

## Author details

Tesfahun Meshesha
Department of Mechanical Engineering, Mizan Tepi University, Tepi, Ethiopia

*Address all correspondence to: tesfahunm.tm@gmail.com

## References

[1] Santamouris M. Minimizing Energy consumption, Energy poverty and global and local climate change in the built environment: Innovating to zero. In: Santamouris M, editor. Energy Consumption and Environmental Quality of the Building Sector. United Kingdom: Elsevier; 2019. pp. 29-64

[2] Zhao X, Yan AZ. Analysis of Energy conservation big data of embedded large public buildings and construction of the information model by 5G. Wireless Communications and Mobile Computing. 2022;**2022**:1-12

[3] European Commission. Commission Recommendation (EU) 2016/1318 of 29 July 2016 on guidelines for the promotion of nearly zero-energy buildings and best practices to ensure that, by 2020, all new buildings are nearly zero-energy buildings. Official Journal of the European Union, Brussels, 2016

[4] Samuel FB, et al. Quadrennial Technology Review: An Assesment of Energy Technologies and Research Opportunities. Washington DC: U.S Department of Energy; 2015

[5] Kent P, Paul T, Roger G. A Common Definition for Zero Energy Building. Energy Efficiency and Renewable Energy Publication and Product Library, 1000 Independence Avenue, SW Washington, DC 20585. 2015

[6] Ditte Juul Jorgensen, "An official website of the European Union," 2023. [Online]. Available from: https://commission.europa.eu/index_en. [Accessed: March 25, 2023]

[7] Delia D'Agostino PZBCDPPB. Synthesis Report on the National Plans for Nearly Zero Energy. Italy: Joint Research Center Science Hu; 2016

[8] United Nations. 2022 Global status report for buildings and construction: Towards a Zero emission, efficient and resilient buildings and construction sector. United Nations Environment Programme, United Nation Avenue. Gigiri Nairobi, Kenya; 2022

[9] Mengxue Lu, Joseph HKL. Building energy: A review on consumptions, policies, rating. 10th International Conference on Applied Energy (ICAE2018), Hong Kong, China, Elsevier; 2019

[10] McDowall R. Fundamentals of HVAC Systems. Oxford, UK, Amsterdam, The Netherlands: Elsevier; 2006

[11] Wilberforce AOESEAT. A review on zero energy buildings – Pros and cons. Energy and Built Environment. 2021:25-38

[12] ASHRAE Standard Project Committee. Energy Standard for Buildings Except Low-Rise Residential Buildings - ASHRAE Standards 90.1. ASHRAE, 1791 Tullie Circle NE, Atlanta, GA 30329. 2021

[13] https://up.codes, 2016. [Online]. Available from: https://up.codes/tmp/viewer_export/juris_key/oregon/pub/ashrae_901_2016/ref/8

[14] U.S. Department of Energy, "Energy Efficiency and Renewable Energy," [Online]. Available from: https://www.energycodes.gov/. [Accessed: March 18, 2023]

[15] I. C. Council, "https://codes.iccsafe.org/content/CEC2019P1/subchapter-

6-nonresidential-high-rise-residential-and-hotel-motel-occupancies-additions-alterations-and-repairs," 2019. [Online]. Available from: https://codes.iccsafe.org/. [Accessed: April 06, 2023]

[16] Jian Z, et.al. Preliminary Energy Savings Analysis: ANSI/ASHRAE/IES Standard 90.1-2019. Pacific Northwest National Laboratory, Richland, Washington 99352

[17] Kelechava B. "American National Standard Institute," 2019. [Online]. Available from: https://blog.ansi.org/ [Accessed: April 06, 2023]

[18] "Wikipedia," 2023. [Online]. Available from: https://en.m.wikipedia.org. [Accessed: March 20, 2023]

[19] Qi X, Saffa R, Shihao Z. Review of Heat Recovery Technologies for building application. Energies; 2019;**12**(7)

[20] Fehrm M, Reiner W, Ungemach M. Exhaust air heat recovery in buildings recovery of heat from extracted air to heat homes. International Journal of Refrigeration. 2002;**25**(4):439-449

[21] Feldman DBDHD. Development and application of organic phase change mixtures in thermal storage gypsum wallboard. Solar Energy Materials and Solar Cells. 1995;**36**(2):147-157

[22] Vali A, Simonson CJ, Besant RW, Mahmood G. Numerical model and effectiveness correlations for a run-around heat recovery system with combined counter and cross flow exchangers. International Journal of Heat and Mass Transfer. 2009;**52**(25-26):5827-5840

[23] Energy Star. Checklist of Common Energy-Saving Measures. United States Environmental Protection Agency, Pennsylvania Avenue NW, Washington, DC 20004, United States, April 2021

[24] U.S Department of Energy. Available from: https://www.energysage.com/energy-efficiency/101/ways-to-save-energy/, 2022. [Online]. Available from: https://www.energysage.com/

[25] "Conserve Energy Future," 2023. [Online]. Available: https://www.conserve-energy-future.com/ [Accessed: March 27, 2023]

[26] Gopal Mishra, "THE CONSTRUCTOR, Building ideas," 2021. [Online]. Available from: https://theconstructor.org/

[27] U.S. Department of Energy, "Tips on Saving Money and Energy at Home," 2022

[28] Elliot Gina, "Top Energy Conservation Tips for Commercial Buildings," 2020. [Online]. Available from: https://www.buildingsiot.com/blog/top-energy-conservation-tips-for-commercial-buildings-bd

[29] "Conserve energy future," 2023. [Online]. Available: https://www.conserve-energy-future.com/. [Accessed: March 29, 2023]

[30] Dekoster USJ. Cycling: The Way Ahead for Towns and Cities. Luxembourg: Office for Official Publications of the European Communities; 1999

[31] "AttuneIOT." 2022. [Online]. Available from: https://www.attuneiot.com/

[32] Erin Scottberg, "This Old House," 2023. [Online]. Available from: https://www.thisoldhouse.com/

[33] "Survey Hub," 15 March 2018. [Online]. Available from: https://www.surveyhub.co.uk/. [Accessed: March 29, 2023]

[34] "Wikipedia," https://en.wikipedia.org/wiki/Solar_energy, 2023. [Online]. Available from: https://commons.wikimedia.org/wiki/Commons:Wiki_Loves_Africa_2023/en. [Accessed: April 06, 2023]

Chapter 4

# Energy-Efficiency Measures to Achieve Zero Energy Buildings in Tropical and Humid Climates

*Katherine Chung-Camargo, Jinela González, Thasnee Solano, Olga Yuil, Vivian Velarde and Miguel Chen Austin*

## Abstract

Nearly and net zero energy buildings have been strongly studied in the global north, with generally a temperate climate, thus focusing on energy-efficiency measures for such climates. Few existing zero energy buildings can be found in tropical and humid climates, where most are ongoing state projects or research projects. Therefore, this chapter brings forth and analyzes the energy-efficiency measures implemented to retrofit buildings in tropical and humid climates to achieve zero-energy buildings. The first part analyzes the measures by energy measure according to building type and climate, envelope (passive and active cooling), energy systems, dampness and mold growth, occupant (thermal comfort and the occupant), renewable energy system (BIPV rooftop and facade BIPV installation), and retrofitting aspects. The second part presents and discusses a proposed framework for policy implementation for building retrofit toward NZEB in Panama based on current building regulations and research evidence, and the viability is assessed via a SWOT analysis.

**Keywords:** energy efficiency, NZEB, policy framework, retrofit, tropical climate, zero energy buildings

## 1. Introduction

Net Zero Energy Buildings (NZEBs) have grown rapidly in developed and temperate countries but face adversities in developing countries, especially under tropical climates. The International Energy Agency (IEA) created the world map of the Solar Heating and Cooling Program (SHC), which presents more than 90% of NZEB projects in developed regions such as the global North. Merely 11 projects are in humid tropical climate zones of developed countries. This led authors such as Feng et al. [1] to consider the economic factor as the more substantial limitation for reaching zero energy buildings in developing regions. As a consequence, they consider that in developing countries, attention to passive techniques and those with relatively low initial investments with short payback periods are critical [1].

Some authors have focused long on studying technologies for mitigating the effects of climate change in buildings, such as zero energy buildings. Up to 2019, different strategies needed in a net-zero-energy building were identified and classified into four categories [2]. The first is related to the design of the building, including ventilation, geometry, and natural light. The second includes energy-saving techniques, contemplating the materials in thermal energy storage systems, envelopes, and energy-efficient household and lighting. The third group comprises renewable energy sources. Finally, backup systems such as electrical energy storage and boilers are considered. The required energy is obtained with the least possible environmental impact [2].

The earliest study on the subject arose in the United States of America. NZEBs were defined as "buildings that obtain enough renewable energy on the site to equal or exceed the annual energy consumption" by The Department of Energy (DOE). Initially, it was proposed to promote building energy efficiency through residential NZEBs in 2020 and commercial NZEBs by 2025. In conjunction with the IEA, the Intergovernmental Panel of Experts on Climate Change (PCC) pointed out that buildings consume 40% of the world's final energy and produce 33% of greenhouse gas emissions [3, 4]. In addition, IEA estimates suggest that the sector's emissions would contribute to an increase of up to 6 degrees of global warming if this industry's activity and trends continue without corrective or mitigation actions [5]. Similarly, between 1971 and 2004, estimated carbon emissions grew by around 2.5% per year in commercial buildings and 1.7% per year in residential buildings, a tendency that continues up to the present [5].

The emission of greenhouse gases (GHG) and the production of waste, together with the use of energy, water, and other resources that are consumed throughout the useful life of the building, are key factors to which different institutions have directed their attention. Some international methodologies and evaluations promote the creation of green buildings through materials, action plans, and measures that significantly reduce environmental impact and optimize the resources used. Particularly noteworthy is the LEED® Certification (Leadership in Energy and Environmental Design), the most popular assessment and qualification system worldwide, which effectively integrates sustainable management of the environment, recycling capacity, prioritization of regional materials, optimal use of water, passive and/or active high-efficiency air-conditioning, and indoor air quality, among other elements [6].

Therefore, this chapter intents to bring forth the energy-efficiency measures implemented in buildings under humid and tropical climates to achieve zero energy buildings and analyze their strengths and opportunities. The first part analyzes the literature to explore past, present, and future research trends regarding the energy-efficiency measures for such climate types. This will be followed by analyzing the measures by type of building and climate, category (envelope, energy systems, mold, occupants), and renewable energy system. Finally, a proposed framework for policy implementation toward ZEB in Panama is explored and evaluated *via* a SWOT analysis.

## 2. Implementation of ZEB in tropical and humid climate

During the last decades, the concept known as NZEB has gained relevance for the construction of buildings and emerged within the different international regulatory frameworks. Very few countries or regions in hot and humid climates have NZEB policies. This lack of guidelines and standards has made NZEB less common in these areas than in developed countries. Therefore, developing countries must concentrate in passive strategies with cost-effective strategies with short payback periods and relatively low initial investing to promote NZEB advance in humid and hot regions, just as policies and incentives play a vital role in promoting NZEB [1]. The energy performance analysis evaluation of buildings should be done on both the macro- and micro-scales, while refraining from doing so could lead to misinterpretation of energy data.

### 2.1 Energy measures according to building type and climate

Continued urbanization and rising income levels of residents in tropical developing countries will require improvements in the indoor thermal comfort of buildings and increased cooling energy demands. Buildings in tropical areas are often designed to improve interactions between indoor and outdoor climates, especially during seasons without air-conditioning. Therefore, an NZEB design is important to use, including passive features such as lighting and natural ventilation, and decrease the building's energy use [1]. Some saving strategies and renewable systems by building type of ZEB in tropical climates are presented in **Table 1**.

In the literature, several methods are mentioned to achieve NZEB buildings, but the following strategies can be summarized for residences [20–23]: (i) propose for comfort and function; (ii) determine an airtight building enclosure; (iii) plan controlled ventilation; (iv) include insulation that surpasses present energy code requirements; (v) certify the building enclosure controls water and moisture movement; (vi) position the building to maximize renewable energy production; (vii) choose efficient mechanical equipment; (viii) choose efficient lighting, appliances, and plumbing fixtures; (ix) energy modeling implementation to predict total energy use, identify high-value improvements to energy efficiency, and size on-site renewables; and (x) elaborate project plans that coordinate and commission systems.

However, energy efficiency measures for buildings with NZEB category are divided into two categories: (1) reduction of the load on the building and (2) more efficient compliance with the load [23].

It can be said that energy efficiency is the first objective of NZEB development, followed by the production of renewable energy from on-site or off-site sources, requirements that cannot be minimized; therefore, to achieve renewable energy technologies such as photovoltaics, wind turbines, solar thermal, district heating, heat pumps, and cooling must be mounted to meet the lasting energy requirements. Likewise, the energy-efficient building envelope is composed of wall thickness, thermal insulation materials, appropriate glazing type, and insulation thickness, which can provide energy efficiency by reducing solar heat increase and cooling load in the subtropical climate in terms of direct solar heat increase, and daylight implementation can be optimized by proper position [24, 25].

### 2.2 Envelope, energy systems, mold, and occupant

The IEA and the US Energy Information Administration presented data that energy consumed by a building is mostly influenced by its ventilation, heating, and air-conditioning (HVAC) [26]. The thermal energy from the environment is one factor that impacts the energy used by the HVAC system in providing occupant comfort [27]. Energy consumption in the domestic sector is governed by several factors,

| Country | Building type | Energy saving strategies | Renewable system | Ref. |
|---|---|---|---|---|
| Brazil | Office | Window systems | BIPV-windows on facade | [7] |
| Brazil | Residential | Advance envelope, optimized shape, and orientation | BAPV-Rooftop | [8] |
| Colombia | Residential | Building orientation, wall insulation, air tightness, window efficiency, and ventilation | PV | [9] |
| Cuba | Hotels | Natural ventilation, lighting | PV | [10] |
| El Salvador | Laboratory | Lighting control | PV | [11] |
| Haiti | Hospital | Light colored walls promote natural ventilation, and reflect solar radiation, operable windows | PV | [12] |
| Panama | Residential | Passive cooling, ventilation, Daylighting control | PV | [13] |
| Panama | University | Passive cooling, envelope modifications, occupant behavior, and building orientation | PV | [14] |
| Malaysia | Commercial | — | BIPV on façade | [15] |
| Sri Lanka | Commercial offices | Shading strategies | BAPV shading strategies | [16] |
| Cameroon | Residential | Thermal insulation in the envelope | BIPV-rooftop | [17] |
| Tanzania | Office | Use of natural light | BIPV-window on façade | [18] |
| Malaysia | Residential | — | BAPV-rooftop | [19] |

**Table 1.**
*Summary of saving strategies and renewable systems by building type of ZEB in tropical climates.*

including the characteristics of the building envelope and consumer behavior, which is why the only option to achieve the required thermal comfort will be the appropriate design of the building envelope [28]. Thus, the first step to achieve thermal comfort in hot climates will always be to avoid the heat, and for this, it is essential to determine what type of cooling is going to be used in the building since a building that uses air-conditioning most of the time must be designed differently than a passive-cooling building [29].

### *2.2.1 Envelope for passive cooling*

Regarding passive cooling, avoiding heat can be achieved with extensive use of shading, white ceilings, light-colored walls, and a well-insulated ceiling. Suppose we focus on passive cooling in hot and humid climates. In that case, the strategies applied are specific to these regions, given the environmental conditions, so ventilation is not intended to cool the thermal mass of a building, but to cool people through evaporation from the skin [29]. Geometry, orientation, and location affect the availability of natural ventilation within the building, especially wind-driven ventilation, which is why a free design is favorable, as it can effectively drive airflow through the building [30]. That is why good air speed in hot and humid climates ensures a comfortable living situation due to humidity and air exchange; wide openings for the wind to enter are essential [31].

There are buildings that can handle both active and passive designs. Such is the case of the University of Brasilia [32]. A laboratory cataloged as NZEB was designed through simulations in DesignBuilder (implementing the loads and schedules suggested by ASHRAE). It was considered an energy saving by allowing natural light, natural ventilation, shade elements, insulation in the ceiling, and an optimal window-wall ratio. In addition, the thickness of the floor is lower to avoid a great absorption of natural light and the efficient use of the air-conditioning system. However, to maintain a balance and not have a greater incidence of sunlight in the enclosure, the orientation of the largest facades is toward the north and south. In its results, evaluated by the indicator Energy use intensity (EUI), it was determined that the balance obtained of 34.30 $kWh/m^2y$ is equivalent to 26% of the average EUI in the city of Brasilia. In addition, it is categorized as a building with a positive energy balance since the on-site generation by the solar panels is greater than the energy consumption of this building. Some countries have applied envelope modifications, such as cooling skins, to provide shade and, in some cases, to generate electricity through photovoltaic modules. As a strategy to improve natural light, ducts were included in the structure of the building in charge of collecting overhead light from the roof and other areas of the facade and redirecting it to spaces where necessary, thus achieving the status of net zero energy building. The complete renovation of this building costs USD 7,639,000 [33].

The context for the strategies of passive cooling usually falls into three categories: (i) reducing heat gains, (ii) changing heat gains, and (iii) removing internal heat [34]. Some techniques applied are designing wind into buildings, using evaporative cooling, double-skin facades, radiant cooling, large windows, solar chimney, and earth cooling [35].

### *2.2.2 Envelope for active cooling*

In tropical/subtropical climates, a lot of energy is consumed for refrigeration, and conventional air conditioners are widely used [24]; envelope systems for buildings in hot and humid climates are the main shield to reduce cooling loads for NZEBs. Those characteristics of the envelopes shared by NZEBs can be summarized in the following parameters [1]:

- Have better thermal integrity to reduce heat transfer and heat gain.
- Reflective roof surface.
- Ventilation layer between the envelope and the interior: double ceiling, attic, or double-skin wall to reduce thermal gain.
- Use of phase change materials in the envelope: to dampen heat transfer.
- Use of vegetation: green walls or roofs manage to reduce heat by approximately 80% and reduce consumption by 2.2 and 16.7%.

- Insulation in exterior windows.
- Shading on windows.

The envelope's U value must stay respectively low by installing insulation layers. This is crucial in buildings with air-conditioning to preserve the cooling load low, mainly on the roof, to reduce the impact of solar radiation. Besides, a continuous layer of insulation surrounding the entire building solves the problem of thermal bridges. In the envelope in hot and humid regions, it is recommended that the window-floor ratio is 15 to 20% as the windows are the main source of gain and need for heat [31].

In modern buildings, the wall's thermal mass has been reduced to save materials, time, and transportation. Light buildings experience large temperature variations, which has led to the incorporation of facades, hydronic systems, thermal conditioning by radiation, thermal storage devices, and many more to reduce energy consumption. However, recent development in building materials has changed this phenomenon with the significant use of phase change materials in hot climates. PCMs may have a significant effect as they have a high volumetric heat capacity during phase change, reaching values about 30 times greater than concrete or conventional building materials. The best measures to test the PCM thermal performance in buildings are melting-point temperature, thermal conductivity, and thermal energy storage density (TES) [36]. A phase change temperature of 27°C performed best in a tropical climate for either residential or office buildings [37].

Considering the investment costs in the economic context of the construction sector in Ecuador, the return periods were estimated in the case of implementing the measures of natural light control and obtained the lowest return period, this being 2.34 years. On the one hand, according to the study, three-layer low-emissivity glazing would have the highest payback period of 103.8 years; on the other hand, replacing the luminaires with more efficient technology resulted in energy savings of 1900 kWh per year and a payback period of 1.2 years [38].

### *2.2.3 Dampness and mold growth*

Generally, the obligations of NZEBs are energy efficiency by reducing envelope thermal transmittance, infiltration, and moisture control. Reduction of heat losses is achieved through the airtightness of the building envelope, which standards can be quite strict [39], by minimizing uncontrolled air movement through infiltration. When adequate ventilation fails to be provided, inadequate air change rates decrease indoor air quality and increase the risk of dampness, surfaces condensation, and fungal growth [39, 40].

This risk could be avoided in the construction process of the building; it is important to ensure that all materials, structures, and products are protected against moisture and weather conditions, while all surfaces need to be dry and water-free before being covered [41]. If not considered during construction, active measures can be implemented for existing buildings during the operation phase. Such measures include air dehumidifiers and increasing ventilation, among others. However, these active measures can increase the overall energy consumption of the building (beyond the PV system generation), therefore resulting in noncompliance with the energy-efficiency regulation requirements for NZEB. On the contrary, when avoiding mold growth risk during construction, only the initial investment costs and the periodic commissioning for indoor environmental quality assurance affect the energy efficiency requirements.

### *2.2.4 Thermal comfort and the occupant*

Zero energy buildings must consider interior comfort conditions (thermal, visual, acoustic). According to ISO 7730 [42] and ASHRAE 55 [26], thermal comfort is a mental situation that expresses gratification with the thermal environment, that is, acceptable indoor air quality, being that which does not contain known pollutants in harmful concentrations as determined by ANSI/ASHRAE 62.1 [43].

The human body can regulate heat flow, ensuring that the heat loss balances the heat generated to keep the body temperature at approximately 37°C [29].

Under the application of HVAC systems, the way to determine comfort levels is by studying comfort indicators and occupant behavior assessments.

The indicator to evaluate thermal comfort corresponds to the Predicted Mean Vote (PMV), which determines thermal comfort according to a scale ranging from −3 to +3, where this index should be in the range from −0.5 to +0.5 to reach thermal comfort conditions.

The IEA and the Energy in Buildings and Communities (EBC) program identify occupant behavior as one of the six determinants of energy use [44]. The occupant's behavior responds to the stimuli generated according to the perception of comfort. Here, the occupant's behavior regards only actions that the occupant executes to seek comfort and to meet their needs by interacting with the building's systems (refer to the methodology DNAS for more information [45]).

## 2.3 Renewable energy system

Existing buildings in hot and humid climates (including tropical ones) can be retrofitted to NZEB by implementing energy-efficiency strategies and integrating photovoltaic (PV) systems (these systems are the most common renewable energy technology adopted by NZEB) [23]. Tropical areas account for 50% of the world's installed capacity. This capacity comprises 20% of solar energy and 20% of wind and 4% biomass. The remaining 56% corresponds to hydropower. Nevertheless, many connoisseurs in the field do not consider the latter renewable energy sources due to their environmental impact [35]. Tropical regions have the highest solar radiation levels, mainly in the desert area of Africa [36]. Brazil has the highest renewable power capacity, followed by Vietnam and Mexico. The rest of the countries maintain a similarity between them. Most energy-efficiency strategies significantly save the building's annual energy consumption and prevent $CO_2$ emissions [46].

PV systems are classified as on-grid and off-grid, where the former refers to a system that is connected to the power grid, while the latter corresponds to systems that are disconnected from the grid; that is, it is an independent system. On the other hand, they can also be classified as on-site, which indicates that the system is installed in the area where the energy produced will be consumed, and off-site is when the installation of the PV systems is in another area that is different from the one where the energy produced will be used [47].

### *2.3.1 BIPV rooftop installations*

The installations depend on the available space in the roof, its inclination, orientation, and bearing capacity. The latter is because the rooftop must support the weight of the panels [48]. Such panels can be curved or flat [49], where curved panels have greater efficiency compared to flat panels in the summer season [42]. On the other hand, this PV design is most in line with the house's architecture and does not compromise greatly with its design [50].

### *2.3.2 Facade BIPV installations*

Over time, the photovoltaic systems installed on the facade will take more prominence because buildings are getting taller. Therefore, the area of the facades is greater than the area of the rooftop [51]. This system can be claddings (BAPV) and curtain walls (BIPV) [52]. If buildings include plants for green designs, these must be coordinated with photovoltaic designs to avoid shadowing from overlapping. This interference must also be avoided by the solar panels themselves (as in other types of installation), which is why the distance-to-length ratio (D/L) is evaluated, where D is the distance between panels and L corresponds to its length. Southeast and southwest facades are the most appropriate for installing photovoltaic panels at 60° as a horizontal tilt angle [45, 46]. However, it has been shown that the horizontal position of the panels on the facade at 45° (instead of 60°) improves efficiency by 8% compared to a vertical position. Considering the distance-to-length ratio for the possible shadows between panels, this ratio should be 3. Despite being very innovative, this technology continues to face difficulties in being as efficient as the panels installed in horizontal planes (rooftop or solar panel ground mounting systems) [46].

Moreover, an application in which solar panels are integrated into buildings employs transparent panels that function as windows, and in this way, it is possible to supply consumption through clean energy; it also allows the entry of natural lighting, reducing the energy demand product of lighting [47]. Semitransparent photovoltaic windows are a good application; however, it is necessary to be careful because of high temperatures, and measures should be established to mitigate this problem [48].

Finally, the return on investment is a very important factor to analyze, whereas the facilities in the tropics present a return in less time due to the high incidence of the sun [49].

## **2.4 Retrofitting aspects**

Unlike in traditional buildings, the operation stage is not responsible for most of the adverse environmental impact in NZEB buildings. The modifications (or retrofitting) cause an increase in the embodied carbon of the materials introduced into the building, so it is necessary to study variables such as the materials manufacturing technologies to determine whether the improvements are viable [53]. The environmental impact caused by such materials could be mitigated with friendlier production processes or by recycling [54]. These adaptations and renewable energy sources undoubtedly involve a cost when bringing a building to the NZEB standard. Thus, the cost of such modifications is considered one of the most important factors when making adaptations to achieve NZEB [49, 50].

The materials in the building envelope considerably increase the investment cost, while their contribution to the decrease of energy consumption is insignificant compared to renewable energies with similar initial costs. Thus, materials must be precisely selected, ensuring that the thermal transmittance values (U) practically improve consumption and keep the investment cost minimal [55]. Note that increasing the insulation thickness in walls decreases the cooling load, especially with green walls. Nonetheless, insulation materials represent a large initial cost [56], where modifications are energetically efficient but not economically, especially in humid climates.

A life cycle cost analysis is recommended to overcome such difficulties when selecting materials, contrasted with the primary energy consumption. However, environmental impact is not directly contemplated in such an analysis, and thus, to evaluate the sustainability of the adaptations, environmental impact indicators must be assessed since results may change considerably when not accounted for such impacts [51].

In a study, the contributions of passive and active design features were compared, and it was found that 40–45% of the energy savings were due to the active design, only 5% of the energy savings were related to the passive design, and air-conditioning upgrades accounted for 23–28% of energy savings, among of which lighting upgrades accounted for 12–17%. The study indicates that passive design should be carefully selected in buildings' retrofit to attain effective results by considering its cost and performance due to different factors such as climate [57].

According to a project, a comparison was made, through simulation, of an existing residence and a version adapted to zero energy through improvements in the envelope and keeping the active design. A 36% reduction in energy demand was determined [9]. In a similar study, it was proven by simulation that a residence decreases between 48 and 50% of its energy consumption by implementing an architectural improvement, decreasing the U-Value of ceiling, floor, and walls to 0.36, 1.83, and 3.293 W/$m^2$K, respectively. In addition, this allowed the installed solar panels to meet all the demand, thus having a positive energy balance [14]. Improvements can also be made to buildings already designed as ZEB, as control systems can be adapted to them. Such is the study [19] where in an office building defined as NZEB, a Model Predictive Control (MPC) is incorporated that works together with a Building Management System (BMS). In this way, you can foresee future problems, optimize the system, and centralize the control for its management. The building had photovoltaic panels, and this energy consumption improved by 19.5%.

A systematic methodology was developed, including the aforementioned energy-efficiency strategies. They outlined previous studies investigating and evaluating energy performance and the economic viability of different retrofit technologies. The authors categorize the energy-efficiency strategies into four groups: (i) improvements in the electrical system and reduction in heating and cooling consumption, (ii) human factors, (iii) energy-efficient equipment and energy reduction technologies, and (iv) renewable energy systems [58].

After this analysis, a summary of the challenges that such types of climates face in implementing energy measures is presented in **Figure 1**. Such challenges can be classified in the availability of climate-specific technological advancements, local capacity buildings, and public awareness campaigns.

The climate-specific technologies focused on moisture control and reducing cooling loads are elements to be carefully considered in tropical climates [59, 60]. Among the local capacity building, the existing building orientation presents a limitation for the feasible measures applicable. Building orientation is a critical parameter for managing heat gains [61]; specifically, this parameter is among the significant predictors for cooling loads in tropical climates [62]. Moreover, double-glazed windows proved to be the most sustainable type for buildings, followed by plenum window type when evaluating their performance in tropical climates, based on energy consumption during building operation, global warming potential emission, embodied energy, and cost [59].

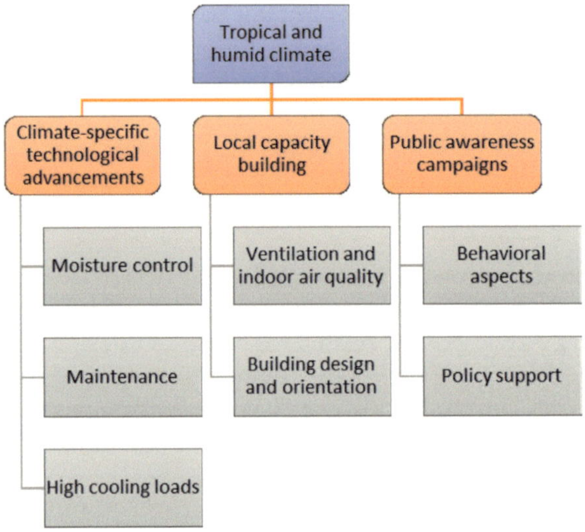


**Figure 1.**
*A summary of the challenges that-energy efficiency measures in tropical and humid climates face.*

## 3. Policymaking framework for buildings retrofit toward NZEB in Panama

In Panama, the building sector accounts for 60% of national electricity consumption (commercial and residential buildings and private and public offices). Passive measures have been observed to help decrease buildings' energy consumption, according to the National Energy Secretary (SNE in Spanish) [63] and the National Energy Plan (PEN in Spanish) [64]. The air conditioner is one of the major energy end-use consumptions in the public and residential sectors. In dwellings, in the case of Panama, 34% is for air conditioners and 26% for refrigerators.

In 2013, through the SNE, the Panamanian government asked for support from the International Finance Corporation (IFC) to develop a Sustainable Construction Guide for energy saving in buildings to mitigate environmental impacts. Consequently, in 2016, the Sustainable Construction Guide [65–67] was established as a regulatory framework to reduce energy consumption in new buildings. Such a framework presents the definition of the average energy consumption according to the type of building: The Baseline. Here, new energy consumption rate requirements for each type of building in Panama are given, along with recommended transmittance values (U-value) (for more details, refer to [65]). In 2019, due to the difficulty in acceptance by the public, this guide was made mandatory through the Sustainable Building Regulation (or RES in Spanish) version 1.0 for energy savings, a simplification of the previous guide. As of early 2023, Version 1.0 of the regulation was revised, updated, and officialized into Version 2.02022 though the energy baseline and savings were kept the same as the initial version [68].

Nevertheless, such regulation applied only to new buildings. In this matter, in 2017, efforts have been jointing with the Green Building Council Panama section and the GreenA Consultants Pte Ltd. of Singapore to develop a voluntary guide for building renovation in Panama, resulting in the so-called Eco Protocol [69]. This is evidence of the lack of local researchers and stakeholders to contribute to local regulations. Current efforts are being made supported by the National Secretary of Science, Technology, and Innovation (SENACYT) through the National Strategic Plan for Science, Technology, and Innovation (PENCYT) 2019–2024. In this plan, the energy sector's challenges are prioritized, such as strengthening the scarce R&D efforts to promote energy saving and transforming the energy matrix (30% from renewable sources) [70]. In this view, the country has taken measures through actions and specific projects of organizations such as the Sectoral Technical Committee, the Ministry of Commerce and Industries (MICI), and the National Secretariat of Energy (SNE), recognizing the necessity to improve energy.

Since 2015, country members of the United Nations have agreed to seventeen Sustainable Development Goals that require policymaking commitments at environmental and sustainable levels. Aiming to provide efficient guidance for this important process, the United Nations for the Environment has established three key principles for considering when creating public policies [71]:

1. Good governance: informed decision-making based on the latest evidence.

2. Health, wealth, and well-being for all: social infrastructures and minorities-oriented

3. Maintain or enhance the environment and natural resources: The background is the center of everything and special care for natural resources.

To achieve more sustainable cities and communities as established in the Sustainable Development Goals, public policies shall consider the different stakeholders of the country's development, engage in sustainable methods, and promote inter-agency coordination for consistent implementation of the regulations [72].

The policymaking process shall be formed by the preparation, public scrutiny, adoption, and publication stages [73]. During the policy preparation, presenting the draft bill to the interested parties and experts is important to obtain feedback that could help enhance the document, considering different perspectives. Furthermore, public institutions and government officials shall also be considered in this scrutinization process, as interinstitutional cooperation is required for creating policies that could be well regulated, implemented, and developed by all offices across the government.

Another common practice in policymaking is the committees and commissions created for debating ideas, either internal or open legislative meetings, with deputies or extended to interested private parties. Participation from the public is necessary as the policymaking process requires the involvement of interested stakeholders that could contribute with knowledge, experience, and best practices. To ensure effective policymaking, citizens must be engaged and called to participate in their creation.

Following up on the preparation of the policy, mechanisms, such as benefits and tax incentives, are often included in public policies related to the environmental sector, pursuing the promotion of investments in the country. It is important to ensure security for private investment through a robust legal framework. Also, since tax incentives are mostly economics-related, the Ministry of Economy and Finance shall be consulted to determine their impact and mitigating initiatives. In this sense, there is literature that highlights the important role of public institutions, which are responsible for the national economy [71].

Furthermore, due to their cross-border significance, international regulations and guidelines are highly considered in creating legal frameworks, especially for the sustainability, environmental, and energy sectors. Sustainable Development Goals, considering sustainable cities and communities, shall be assessed in creating public policies for net zero energy buildings [74].

### 3.1 Panamanian legal framework and structure

For the adoption and publication of the policy, Panamanian regulations establish three rounds of legislative debates (**Figure 2**). For the first debate session, public participation is permitted by listeners. Moreover, legislators could require modifications to the policy in the review during the first and second debate sessions. It is important to understand that no further changes are allowed for the third debate session, and its exclusive purpose is the overall discussion of the bill of policy.

The document is sent to the Executive for presidential sanction upon legislators' approval of the bill of policy. In this stage, the president could either object to it, approve/sanction it, or object due to unconstitutionality. If the bill of policy is approved, it will be the policy of the Republic during the following 6 days. When the president objects to a bill of policy, the document is returned to the legislators for review and continuation of its application process.

Discussing the legal framework for energy-related projects, it is important to understand the political structure of the public authorities in Panama. In this sense, by means of the Policy N° 43 of April 25, 2011, the National Secretary for Energy was reorganized and officially appointed to propose and promote public policies for the energy sector and ensure the security of supply and rational and efficient use of resources and energy in a sustainable manner [75].

On the other hand, the National Authority of Public Services ("ASEP") is the regulatory entity for electricity-related services. As the regulator, ASEP is entitled to create and publish resolutions, guidelines, and regulations that will facilitate the implementation of policies in this sector. Furthermore, public policy creation should consider establishing standards and criteria that the regulator could apply for policy enforcement; this is as important as creating suitable policies. This said, in addition to

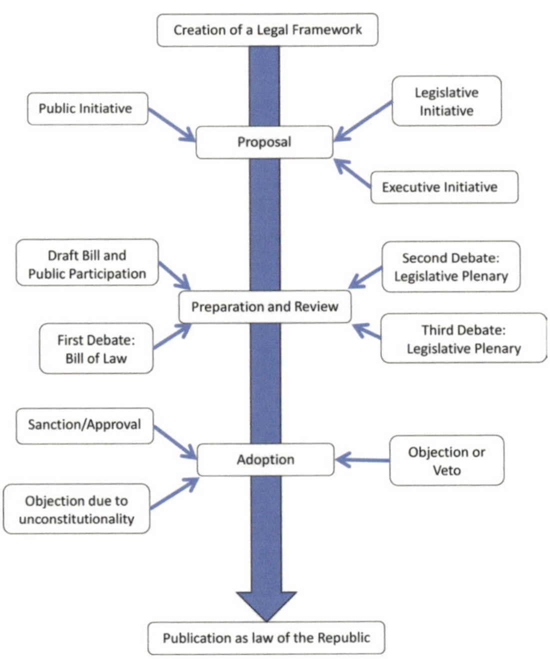


**Figure 2.**
*Creation of a legal framework flow in Panama.*

the legislative, these two government entities are entitled to the policymaking process in the energy sector.

Currently, some regulations regarding sustainable constructions in Panama provide appropriate measures for the rational and efficient use of energy in new constructions. Considering this, the energy demand in those buildings will be lower, and its supply will be easier to achieve [76]. As previously mentioned, the Policy 69 of 2012 recognizes the obligation of new multifamily houses, as well as industrial, commercial, and public constructions, to comply with technical guidelines and regulations for the rational and efficient use of energy.

Resolution N° 3142, dated November 17, 2016, published the Guide for sustainable construction for energy saving in buildings and measures for the rational and efficient use of energy for new constructions in the Republic of Panama. This document establishes the energy consumption baseline according to the construction type, aiming to accomplish the saving percentages. For the review of these guidelines, it establishes a five-year periodicity. Although there is some guidance on efficient construction procedures, there are not yet specific regulations for achieving net zero energy buildings in Panama.

### 3.2 Evidence of buildings retrofit toward NZEB in Panama

On the one hand, the Eco Protocol [69] is the only building retrofit proposal in Panama. It takes as a baseline the same as the RES version 2023 [77] and gives voluntary recommendations regarding envelope energy efficiency, water consumption, and waste management. Although much reference is taken from Singapore regulations, the recommended envelope energy-efficiency values are similar to those in [77].

On the other hand, Panama, as with different types of tropical climates, has some published scientific studies about ZEB. Among these studies, only one resulted in an existing building. However, such a building was designed with NZEB intentions, applying passive and active strategies according to the climate [16] and did not undergo a retrofit procedure. This study implemented passive dehumidification techniques, lighting, natural ventilation, high-efficiency windows, and solar collectors to reduce electrical consumption and mechanical ventilation equipment. To obey the NZEB standard, a photovoltaic system with batteries was installed to supply the demand for the two-story house [16].

Moreover, a test model was developed based on common characteristics from numerical assessments. Dynamic simulations with typical meteorological data helped identify some useful techniques, such as determination of the orientation of the building, natural ventilation, and modifications in the occupancy profile and the envelope [17]. Similarly, another study employing dynamic simulations was performed at an urbanization scale with 34 residences [78]. The original urbanization was redesigned following the bioclimatic principles for tropical climates [25] to reduce the cooling needs, and optimization analyses were performed to determine adequate materials. It was found that cooling could be entirely reduced by increasing the building compacity and implementing only natural ventilation to promote indoor thermal comfort—such a design entails high embedded carbon emissions due to the bigger dimensions [78].

Only two numerical studies exist in Panama regarding the retrofitting of buildings *via* dynamic simulations at the building scale [79] and at urbanization scale [80], using the optimization technique contrasting total construction cost with electricity consumption. At the building scale, preliminary technological solutions to adapt existing buildings in Panama into zero energy buildings were identified by employing sensitivity and optimization analyses combined with total construction costs. Results showed that the critical variables influencing electricity consumption are cooling set point, window-to-wall ratio (WWR), and cooling operation schedule. The most appropriate solutions were: (i) When WWR near 30% is recommended, lower cooling set points can be used (near 24°C); (ii) when a higher WWR than 30% is recommended, a higher cooling set point is required (near 27°C) with high CoP values; and (iii) either the appliances increase efficiency, or the occupants' usage needs to change [79]. At the urbanization scale, after applying optimization techniques within the retrofit category related to the building envelope and occupant behavior, a 60% electricity reduction was attained while reducing the renewable energy generation capacity to 10%. One promising solution is to combine the human factor with minimal changes to the building envelope, as it requires minimal investment. Applying extreme modifications to occupant behavior, which are impractical from many views, was the best option found, but this leads to the same indications reported in several NZEB research where the greater the impact on occupant behavior, the higher the energy performance of the building [80].

All four studies concluded that the energy generated from PV systems was enough to cover the electricity needs of the building.

Although important insights were obtained from such studies, results remain preliminary due to the reproducibility of the results. Based on the optimized solutions encountered here, bioclimatic strategies could substantially impact electricity consumption. As recommended for tropical climates, among the strategies are incorporating local shading and increasing the WWR, uninsulated constructive systems, and window blinds. Finally, for better outcomes, there is the need to develop a building stock inventory with their construction features and perform the optimization methodology on the most representative building for each building type [79].

## 3.3 Further support for retrofit policymaking

### *3.3.1 International certifications in Panama*

This section studies the most used green building programs in Panama: Excellence in Design for Greater Efficiencies (EDGE), and Leadership in Energy and Environmental Design (LEED). Green building programs and certifications propose strategies or efficiency measures as an integral part of the solution to the planet's environmental challenges.

#### *3.3.1.1 Leadership in energy and environmental design (LEED)*

The U.S. Green Building Council's LEED green building program is the outstanding program for designing, constructing, maintaining, and operating high-performance green buildings. LEED is a voluntary and market-driven guideline and assessment mechanism that can address commercial, institutional, and residential buildings and neighborhood developments [81, 82]. The program intends to promote a transformation of the construction industry through goals in climate change, human health, natural resources, a greener economy, and community.

Within LEED, there are four main rating systems; one must be selected according to the project case. (1) LEED for Construction and Building Design: new construction or major renovation in Buildings, (2) LEED for Construction and Interior Design:

Interior spaces that are a complete interior fit-out, (3) LEED for Building Operations and Maintenance: Totally operational and occupied buildings for at least 1 year, and (4) LEED for Neighborhood Development: New land growth projects or redevelopment projects containing nonresidential uses, residential uses, or a mix.

Each rating system has six main categories: Materials and Resources (MR), Location and Transportation (LT), Sustainable Sites (SS), Water Efficiency (WE), Energy and Atmosphere (EA), and Indoor Environmental Quality (EQ). Each category comprises a diverse group of strategies called credits and prerequisites, where points are weighed and allocated based on the relative importance of their contribution to the program's goals.

LEED's EA prerequisite minimum energy performance will be explored in this section. The objective of minimum energy performance is to narrow economical and environmental harms because of excessive energy consumption by achieving the lowest level of energy efficiency in the systems on the proposed building. Minimum energy performance must be established either by developing a whole-building energy simulation or by following a prescriptive compliance, in both options according to ASHRAE-90.12010 requirements. For the current Version 4 of the LEED program, the referenced standard for building the baseline model was updated from ASHRAE 90.1–2007 to ASHRAE 90.1–2010, representing a substantial increase in efficiency.

#### *3.3.1.2 LEED zero program*

LEED Zero spotlights the performance of exemplary projects in topics that are critical to the goal of sustainable development. Projects must be LEED certified under a Building Operations and Maintenance or Building Design and Construction rating system and can pursue multiple of the following LEED Zero certifications at the same time [83]:

- LEED Zero Carbon: recognizes buildings operating with net zero carbon emissions throughout the past 12-month period. A visible accounting of the balance of carbon caused by energy consumption and occupant transportation to carbon emissions avoided or offset is provided.

- LEED Zero Energy: distinguishes buildings that accomplish a source energy use balance of zero for the past 12-month period.

- LEED Zero Water: distinguishes buildings that accomplish a potable water use balance of zero for the past 12-month period.

- LEED Zero Waste: distinguishes buildings that accomplish Zero Waste certification at the highest level and are verified by a third-party organization.

Certifications are valid for 3 years from the date of certification acceptance and must be recertified every 3 years to maintain an updated status.

For the LEED Zero Energy certification, the net zero energy balance is founded on the amount of source energy delivered and the amount of renewable energy that displaces nonrenewable energy on the grid. Renewable energy generated and used on-site reduces the amount of energy delivered.

To calculate source energy delivered to the project, the LEED Zero program indicates to use the national ENERGY STAR Source-Site Ratios for each building energy source from the Energy Star Portfolio Manager Technical Reference: International projects may use the U.S. source-to-site ratios or issued source-to-site ratios for the country or multicountry region where the project is located and Source Energy for projects in the U.S. and Canada [84]. Equal source energy conversion aspects can calculate the energy delivered and nonrenewable energy displaced.

For energy balance, there is also the option of purchasing Renewable Energy Certificates (RECs), which must be Green-e Energy certified or equivalent.

#### *3.3.1.3 Excellence in design for greater efficiencies (EDGE)*

EDGE is a green buildings platform that incorporates a green-building average, certification program, and a software application. EDGE is a member of the World Bank Group and innovation of IFC. The building typologies supported by this standard include houses, apartments, hospitals, offices, hotels, retail, and education buildings. The standard intends to project operational savings and reduce environmental impacts such as carbon emissions [85], using a simple and cost-effective interactive tool that can be used to plan and evaluate the design of resource efficiency to potentiate green building growth and market. The interface is promoted to constantly evolve as an easy-to-use tool and avoid the need to hire energy specialists or purchase modeling software.

A building is awarded EDGE certification when the required minimum efficiencies of 20% are achieved specifically for the following three categories: energy, water, and materials. The complete certification process is conducted online through software called EDGE App. Compared to a base case model, the App demonstrates a minimum of 20% savings in operational water, energy, and embodied energy in materials in the analyzed building. Certification is awarded in three statuses according to the project efficiencies: (1) EDGE Certified: after achieving 20% efficiencies in energy, water, and materials each; (2) EDGE Advanced: when the project has achieved 40% or greater savings in energy, beyond the minimum 20% required, and (3) EDGE Zero Carbon: awarded to projects that demonstrate zero carbon emissions in operations either through renewables or through carbon offsets; an EDGE Advanced status is required previously [86].

The EDGE water and material efficiency measures are recommended for further analysis and future studies.

EDGE base case, improved case (analyzed building), and savings projections are calculated based on three categories: (1) climatic conditions of the location, (2) building type and occupant use, and (3) design and specifications [86]. As part of the standard's methodology, if there is a sustainable building regulation in practice in a certain country, it is used to develop the base case calculations. For Panama, it is the sustainable construction guidelines. As for heating, ventilation, and air-conditioning typical system efficiencies, the current EDGE Version 3.0.a (corresponds to EDGE Software Version 3.0) is based on ASHRAE-90.12007. This might fall behind compared to LEED Version 4, which is based on a more recent version of ASHRAE.

### *3.3.2 Building stock inventory and costs*

When planning for a massive retrofitting procedure, proposing passive and active renovation strategies by focusing on an individual building become impractical because each building has its characteristics. Therefore, such passive and active strategies should be proposed over a typical building that represents most of the buildings of a specific type. This can be addressed by carrying out a building stock inventory, which statistically allows us to find the most frequent characteristics regarding construction materials (for floors, walls, partitions, and roof), year of construction, surface area, number of rooms, number of stories, zones distribution, household, energy use, systems (ventilation, air conditioning), and energy consumption.

Moreover, the investment costs involved in the renovation process should also be considered, for example, replacement of insulation layers, roof replacement, changing of windows, upgrading the air-conditioning systems, and demolition, among others. Current mandatory sustainable building regulations in Panama do not account for the investment costs involved in the energy efficiency requirements for new buildings; that is, recommendations are not based on energy-efficiency improvement contrasted with the investment cost.

Unfortunately, no building stock inventory is available that considers those mentioned above. In contrast, scarce information can be found in different separate studies conducted. **Table 2** shows the average house area and occupancy for apartments and houses base case, classified by income categories. Similar averages were presented and studied [88] regarding four house size ranges: 60–80, 80–100, 100–120, and 120–150 $m^2$. Thus, the need for a building stock inventory is urgent.

## 4. Discussion

Analyzing the current local regulations and retrofit evidence from recent studies in Panama, two scenarios withstand regarding the objective of buildings undergoing a retrofit procedure:

1. Renovate buildings to increase energy efficiency to comply with current regulations, that is, RES [77] (SC1).

2. Renovate buildings to increase energy efficiency to comply with NZEB (SC2).

However, more data is needed to choose between both scenarios. This highlights the importance of the building stock inventory. Current data available in Refs. [78, 80, 88] suggests the following highlights at the residential level:

| | Income category | No. of bedrooms | Average house area ($m^2$) | Occupancy (people/house) |
|---|---|---|---|---|
| | | Houses | | |
| 1 | Low | 1 | 60 | 3 |
| 2 | Middle | 3 | 145 | 4 |
| 3 | High | 5 | 250 | 6 |
| | | Apartments | | |
| 1 | Low | 1 | 60 | 3 |
| 2 | Middle | 3 | 120 | 4 |
| 3 | High | 5 | 200 | 6 |

**Table 2.**
*Standard dimensions for houses and apartments in Panama's case base in EDGE app [87].*

1. For the thirty-four houses analyzed, an average of 0.44 $kWh/m^2$/month was encountered for its reference case per house, which yields 1.4 $kWh/m^2$/month in terms of primary energy [78], without considering gas consumption.

2. For the thirty-eight houses, an average of 1.90 $kWh/m^2$/month was found per house (or 6.0 $kWh/m^2$/month) in primary energy terms [89] without considering gas consumption.

3. Assuming the RES [77] does not consider gas consumption in its baseline, an average of 293.4 $kWh/m^2$/y yields if gas is considered for a 60 $m^2$ house in terms of primary energy (24.45 $kWh/m^2$/ month, with 3.15 and 1.15 factors, for electricity and gas, respectively). This would yield 252 $kWh/m^2$/y (or 21 $kWh/m^2$/month) for electricity only in primary energy terms. Considering the 15% saving proposal of [77], this yields 17.86 $kWh/m^2$/month in primary energy.

4. Implementing only modifications to the envelope does not yield a large reduction in energy consumption as changes in occupants' habits. This could yield up to a 60% reduction.

Further discussion of this is presented hereafter by highlighting the strengths, weaknesses, opportunities, and threads in applying ZEB in Panama, based on and adapted from experiences reported in [52, 79–86, 88].

Among the strengths is improving the engineering, architectural, and construction industries and maintaining leadership for high-performance building in Latin America, through the potential use of energy-efficient systems and using renewable energy sources. This, in turn, allows the reduction of operational energy costs and helps to mitigate the reduction of energy resources and the decline of our environment. In addition, the NZEB application can improve comfort terms and become a realistic solution for mitigating $CO_2$ emissions and/or reducing energy use in the building sector. Many countries have already established zero energy buildings as their future building energy target, which increases their role as a promising potential to significantly reduce energy use and increase the overall share of renewable energy.

An end-to-end structure supporting drafting, revising, and publishing the bill of laws could be mentioned among the strengths of the policymaking process in Panama. In this sense, it is important to highlight that the Panamanian process is quite expedited, with only three stages.

Another strength is the diversity of legal guidance about energy efficiency in Panama, considering this as an appropriate path for implementing new policies about related matters. When discussing the strengths of the policy implementation in Panama, citizenship participation in the law-making shall be mentioned. In this sense, Panama allows civil participation to draft a Bill of law and present it to the National Assembly for debate. Through this mechanism, independent professionals, academics, and other private sectors are entitled to promote the necessary policies and are granted an opportunity to be reviewed by the legislators.

The opportunities of NZEB systems lie in acceleration to achieve national goals to increase the renewal integration in the building sector. In addition, it establishes a transformation in the built environment by demonstrating NZEB as a pilot project or showcase for energy-neutral buildings and achieves the development of an adaptive comfort model for Residential buildings in Panama. On the other hand, it increases research innovation in this sector by increasing knowledge competence and mitigating the lack of technologies. In addition, it is involved in developing new energy policies, thus allowing for the incorporation of zero energy buildings into Panama's Sustainable Building Regulation. Even this allows the retrofitting and building optimization and proceeds through correctly establishing the associated design techniques and technologies for zero energy buildings. Reducing energy demand is an opportunity posed by NZEB systems. Hence, the greatest margin to do this resides in improving the real estate stock, where the importance of sustainability in energy policy and climate change is concentrated.

Having legal stability and a clear legal background promotes private investment, which is a considerable opportunity for Panama, to the extent that having the necessary resources helps to continue researching and improving an area of interest. Furthermore, job creation comes together with the country's development and benefits all citizens. For the specific case of the energy sector, if Panama pursues a strong legal background and regulations, it is likely to receive support from international bodies and therefore investments for projects in this area. Another area of opportunity is broadening communications about the policymaking process in Panama, fostering the participation of professionals that could provide their expertise.

However, NZEB systems have certain weaknesses in the economy, industrial sector, human resources, and installations due to the design of the building. The first refers to the low dedicated budget for financial aid to construct new buildings and rising costs due to the need for high-level technology such as heat recovery systems, control building materials, and components. The second concerns the lack of industrial infrastructure and dependence on imports for technological components. The third is based on the lack of knowledge, competence, and skill labor for the design and construction of NZEB, besides insufficient institutional and available human resources, and can be added the lack of mastery of technological tools necessary for data analysis, since the need for a robust calculation methodology has gained attention and thus the interest in how the zero balance is computed. The last approach addresses the existing limitation of on-site energy generation options (e.g., appropriate space for solar systems in high-rise buildings).

For mentioning weaknesses of the policy implementation process in Panama, it must be considered that the legislators of the National Assembly are publicly elected positions, being common to find a representation of multiple parties, resulting in cases where a bill of law could have multiple delays due to lack of consent. This becomes relevant to the extent that the lack of political consensus jeopardizes the entire process, slowing it down and causing a disruption. On the downside of the existence of energy efficiency policies, there might be different legal instruments, but a gap continues on the regulatory side, hindering its enforcement.

The threats facing NZEB systems reside in a lack of policies and standards, which results for an unreasonable energy structure, where there is no consensus among educators, families, government, and local authorities, given the complexity in understanding zero energy buildings. NZEB installations face the threat of the structural design of the building, given that the building's embodied energy and thermal insulation have often been overlooked in the green building certification system. On the other hand, there is a slow and weak market transformation toward major recycled and low-embodied building materials. There is also noncompliance with the correct management of processes to guarantee the quality of operation of the NZEB installation. In addition, an absence of energy at the regional level could cause high electricity costs to conserve income balance in the energy sector.

It is well-known that technology tends to develop faster than regulations, which poses a threat to the policymaking process as individuals could work in something that has not yet been regulated or within a gray area. The lack of long-term commitment is another threat to the Panamanian policy implementation process. Considering that the government changes every 5 years, including legislators, even those good legislative initiatives are often neglected after a new political party takes control.

The weaknesses and opportunities appear to back up the pertinence of the ZEB framework in Panama. Most Latin American countries (economic sector and human resources and integrations of all characters) attain an actual situation as in Panama and might be limiting going further in applying ZEB than remaining in a preliminary research status.

Specifically in Panama City, new buildings show higher potential than existing high-rise buildings in the public and commercial sectors. This is mainly due to a significant glazing surface area without openings in high-rise buildings by the seaside, demanding an always-on air conditioning activated system to ensure both air quality permanence and indoor thermal comfort. This is clearly opposed to all precedent studies regarding bioclimatic architecture in tropical climates.

## 5. Conclusion

Energy efficiency measures to achieve zero energy buildings in tropical and humid climates are still trying to define regulations that set minimum and maximum energy efficiency levels for different types of buildings, while in other climates, building energy codes already require buildings to be net-zero energy.

Thus, current research in Panama gathers lower energy consumption values in existing buildings than the baseline and the 15% savings proposal reported in the latest local regulation. This concludes that a tremendous amount of work is still needed. However, such preliminary results highlight the potential toward achieving the second scenario (SC2) by implementing renovation policies for existing buildings (at least for houses) by focusing most on occupants' behavior (air conditioner temperature setpoint and schedule) rather than on building envelope.

Finally, international certifications such as LEED and EDGE can help establish limits for buildings emissions, water consumption, and waste management for existing buildings.

## Acknowledgements

The authors would like to thank the Faculty of Mechanical Engineering (https://fim.utp.ac.pa/) to the Research Group Energy and Comfort in Bioclimatic Buildings (ECEB, https://eceb.utp.ac.pa/) within the Universidad Tecnológica de Panamá (https://utp.ac.pa/) for their cooperation. This investigation was funded by Panamanian Institution Secretaría Nacional de Ciencia, Tecnología e Innovación (SENACYT, https://www.senacyt.gob.pa/) and Sistema Nacional de Investigación (SNI) under the project code IDDS22-30.

## Conflict of interest

The authors declare no conflict of interest.

## Author details

Katherine Chung-Camargo[1], Jinela González[1], Thasnee Solano[1], Olga Yuil[2], Vivian Velarde[3] and Miguel Chen Austin[1*]

1 Research Group Energy and Comfort in Bioclimatic Buildings (ECEB), Faculty of Mechanical Engineering, Technological University of Panama, Panama City, Panama

2 CBRE Panama, Plaza Credicorp Bank, Ciudad de Panamá, Panama

3 Energy Law Commission of the National College of Lawyers, Ciudad de Panamá, Panama

*Address all correspondence to: miguel.chen@utp.ac.pa

## References

[1] Feng W et al. A review of net zero energy buildings in hot and humid climates: Experience learned from 34 case study buildings. Renewable and Sustainable Energy Reviews. 2019;**114**:109303. DOI: 10.1016/j.rser.2019.109303

[2] Cabeza LF, Chàfer M. Technological options and strategies towards zero energy buildings contributing to climate change mitigation: A systematic review. Energy and Buildings. 2020;**219**:110009. DOI: 10.1016/j.enbuild.2020.110009

[3] I. E. Agency. Promoting Energy Efficiency Investments. Paris, France: I. E. Agency; 2008

[4] IPCC. Comunicado de premsa IPCC, Grup. Intergub. Expert. sobre el Cambio Climático. Switzerland: IPCC; 2018. pp. 1-5. Available from: https://www.ipcc.ch/site/assets/uploads/2018/11/pr_181008_P48_spm_es.pdf

[5] Ürge-Vorsatz D, Harvey LDD, Mirasgedis S, Levine MD. Mitigating $CO_2$ emissions from energy use in the world's buildings. Building Research & Information. 2007;**35**(4):379-398. DOI: 10.1080/09613210701325883

[6] Garcia E. Edificios 'Net Zero' la próxima generación de construcciones sustentables. Available from: https://inmobiliare.com/edificios-net-zero-la-proxima-generacion-de-construcciones-sustentables/

[7] Didoné EL. Parametric Study for Net Zero Energy Building Strategies in Brazil Considering Semi-Transparent PV Windows. Germany: Department of Architecture, KIT Department of Architecture; 2014. p. 230. DOI: 10.5445/IR/1000044721

[8] Domingos RMA, Gabriel E, Guarda ELA, Pereira FOR. Assessment of technical-financial analysis of a zero energy building for Brazilian hot and temperate tropical climate. IOP Conference Series: Materials Science and Engineering. 2019;**609**:72013. DOI: 10.1088/1757-899X/609/7/072013

[9] Nematchoua MK et al. Transition to Zero Energy and Low Carbon Emission in Temperate Climates. Switzerland: MDPI; 2021

[10] Akata AMEA, Njomo D, Agrawal B. Assessment of building integrated photovoltaic (BIPV) for sustainable energy performance in tropical regions of Cameroon. Renewable and Sustainable Energy Reviews. 2017;**80**:1138-1152. DOI: 10.1016/J.RSER.2017.05.155

[11] Kachi A, Warnecke C, Hagemann M, Nascimento L, Mooldijk S, Tewar R. Net-Zero Energy Housing Virtual Article 6 Pilot. Vol. no. January. Germany: New Climate Institute; 2020. p. 41

[12] Collado Baldoquin N, Guzmán LAR, Dania GC. Hoteles de Consumo Energético Casi Nulo. In: Potencialidades y Restricciones para Cuba. Cuba: Cujae; 2018. Available from: https://www.researchgate.net/profile/Gonzalez-Couret-Dania/publication/330542847_HOTELES_DE_CONSUMO_ENERGETICO_CASI_NULO_POTENCIALIDADES_Y_RESTRICCIONES_PARA_CUBA/links/5c474b7792851c22a3882308/HOTELES-DE-CONSUMO-ENERGETICO-CASI-NULO-POTENCIALIDADES-Y-RES

[13] Nzeb-El Salvador. Editorial Construir. Available from: https://www.construir.connectab2b.com/post/nzeb-el-salvador-el-edificio-de-cero-energía-neta-de-us-400-000

[14] Ohene E, Hsu SC, Chan APC. Feasibility and retrofit guidelines towards net-zero energy buildings in tropical climates: A case of Ghana. Energy and Buildings. 2022;**269**:112252. DOI: 10.1016/j.enbuild.2022.112252

[15] Guenther G, Vittori R. Sustainable Healthcare Architecture. USA: John Wiley & Sons; 2013

[16] Hoque S, Iqbal N. Building to net zero in the developing world. Buildings. 2015;**5**(1):56-68. DOI: 10.3390/buildings5010056

[17] Mora D, Araúz J, Austin MC. Towards nearly zero energy buildings in Panama through low-consumption techniques: A numerical study. In: AIP Conference Proceedings. 2019;**2191**(December):020114-1-020114-9. DOI: 10.1063/1.5138847

[18] Hamzah AH, Go YI. Design and assessment of building integrated PV (BIPV) system towards net zero energy building for tropical climate. e-Prime – Advances in Electrical Engineering, Electronics and Energy. 2023;**3**:100105. DOI: 10.1016/j.prime.2022.100105

[19] Zhan S, Dong B, Chong A, Zhan S, Dong B, Chong A. Improving energy flexibility and PV self-consumption for a tropical net zero energy office building. Energy and Buildings. 2023;**278**:112606. DOI: 10.1016/j.enbuild.2022.112606

[20] Mendis T, Huang Z, Xu S. Determination of economically optimised building integrated photovoltaic systems for utilisation on facades in the tropical climate: A case study of Colombo, Sri Lanka. Building Simulation. 2020;**13**(1):171-183. DOI: 10.1007/s12273-019-0579-4

[21] Joseph B, Kichonge B, Pogrebnaya T. Semi-transparent building integrated photovoltaic solar glazing: Investigations of electrical and optical performances for window applications in tropical region. Journal of Energy. 2019;**2019**:1-10. DOI: 10.1155/2019/6096481

[22] Lau KY, Tan CW, Ching KY. The implementation of grid-connected, residential rooftop photovoltaic systems under different load scenarios in Malaysia. Journal of Cleaner Production. 2021;**316**(May 2020):128389. DOI: 10.1016/j.jclepro.2021.128389

[23] Wu W, Skye HM. Residential net-zero energy buildings: Review and perspective. Renewable and Sustainable Energy Reviews. 2021;**142**(March 2020):110859. DOI: 10.1016/j.rser.2021.110859

[24] Rana MJ, Hasan MR, Sobuz MHR, Sutan NM. Evaluation of passive design strategies to achieve NZEB in the corporate facilities: The context of Bangladeshi subtropical monsoon climate. International Journal of Building Pathology and Adaptation. 2021;**39**(4):619-654. DOI: 10.1108/IJBPA-05-2020-0037

[25] Olgyay V. Design with Climate: Bioclimatic Approach to Architectural Regionalism. New and expanded ed. USA: Princeton University Press; 2015

[26] American Society of Heating, Refrigerating and Air-Conditioning Engineers. ANSI/ASHRAE Standard 55. Journal of Chemical Information and Modeling. 2017;**53**(9):1689-1699

[27] Chiradeja P, Ngaopitakkul A. Energy and economic analysis of tropical building envelope material in compliance with Thailand's building energy code. Sustainability. 2019;**11**(23):6872. DOI: 10.3390/SU11236872

[28] Pathirana S, Rodrigo A, Halwatura R. Effect of building shape, orientation, window to wall ratios and zones on

energy efficiency and thermal comfort of naturally ventilated houses in tropical climate. International Journal of Energy and Environmental Engineering. 2019;**10**(1):107-120. DOI: 10.1007/S40095-018-0295-3

[29] Lechner N. Heating, Cooling, Lighting: Sustainable Design Methods for Architects. 4th ed. USA: Wiley; 2015

[30] Yu CR, Sen Guo H, Wang QC, Chang RD. Revealing the impacts of passive cooling techniques on building energy performance: A residential case in Hong Kong. Applied Sciences. 2020;**10**(12):4188. DOI: 10.3390/APP10124188

[31] Sudhakar K, Winderla M, Priya SS. Net-zero building designs in hot and humid climates: A state-of-art. Case Studies in Thermal Engineering. 2019;**13**:100400. DOI: 10.1016/j.csite.2019.100400

[32] Amorim CND et al. An integrated design process in practice: A nearly zero energy building at the University of Brasília-Brazil. In: Nearly Zero Energy Build. – Material, Design, and New Approaches. London, UK: IntechOpen; 2022. DOI: 10.5772/intechopen.102443

[33] Wittkopf S. Tropical net zero. In: High Performing Buildings, no. 76 m. USA: High Performing Buildings; 2015. pp. 44-54. Available from: www.hpbmagazine.org

[34] Al-Obaidi KM, Ismail M, Rahman AMA. Passive cooling techniques through reflective and radiative roofs in tropical houses in Southeast Asia: A literature review. Frontiers of Architectural Research. 2014;**3**(3):283-297. DOI: 10.1016/J.FOAR.2014.06.002

[35] Savero RR, Antaryama IGN, Soemardiono B, Antaryama GN. Review on design strategies of energy saving office building with evaporative cooling in tropical region. IPTEK The Journal for Technology and Science. 2020;**31**(2):236-248. DOI: 10.12962/J20882033.V31I2.5607

[36] Sangwan P, Mehdizadeh-Rad H, Ng AWM, Tariq MAUR, Nnachi RC. Performance evaluation of phase change materials to reduce the cooling load of buildings in a tropical climate. Sustainability. 2022;**14**(6):3171. DOI: 10.3390/SU14063171

[37] Austin MC, Araúz J, Mora D. Numerical assessment of different phase change materials as a passive strategy to reduce energy consumption in buildings under tropical climates. Buildings. 2022;**12**(7):906. DOI: 10.3390/BUILDINGS12070906

[38] Mohit OK, Oree V. Assessing the energy savings potential in public buildings through retrofit measures in tropical climates — A case study in Mauritius. In: 2013 Africon. USA: IEEE; 2013. pp. 1-5. DOI: 10.1109/AFRCON.2013.6757745

[39] Di Giuseppe E. Nearly Zero Energy Buildings and Proliferation of Microorganisms: A Current Issue for Highly Insulated and Airtight Building Envelopes, no. 9783319023557. USA: Springer International Publishing; 2013

[40] Brambilla A, Sangiorgio A. Mould growth in energy efficient buildings: Causes, health implications and strategies to mitigate the risk. Renewable and Sustainable Energy Reviews. 2020;**132**:110093. DOI: 10.1016/j.rser.2020.110093

[41] Pihelo P, Kalamees T. Commissioning of moisture safety of nZEB renovation with prefabricated timber frame insulation wall elements. Wood Material Science &

Engineering. 2019;**16**(2):110-117. DOI: 10.1080/17480272.2019.1635206

[42] del Valle FC. Conceptos Generales sobre Ambiente y Confort Térmico. Chávez del Valle, FJ. 2002;**2**:19-36. Available from: http://www.tesisenred.net/bitstream/handle/10803/6104/07CAPITULO2.pdf?sequence=7&isAllowed=y

[43] S. A. de Ingenieros de Calefacción refrigeración y acondicionamiento de aire Inc. ANSI/ASHRAE 62.1 Ventilacion para una Calidad Aceptable de Aire Interior. USA: S. A. de Ingenieros de Calefacción refrigeración y acondicionamiento de aire Inc., Estándar; 2007. p. 12

[44] Agency IE. IEA-EBC: Energy use in buildings. International Journal of Ambient Energy. 2015;**36**(2):49. DOI: 10.1080/01430750.2015.1013007

[45] Hong T, D'Oca S, Taylor-Lange SC, Turner WJN, Chen Y, Corgnati SP. An ontology to represent energy- related occupant behavior in buildings part I: Introduction to the DNAs framework. Building and Environment. 2015;**94**(P1):196-205. DOI: 10.1016/j.buildenv.2015.08.006

[46] Alajmi A, Short A, Ferguson J, Vander Poel K, Griffin C. Detailed energy efficiency strategies for converting an existing office building to NZEB: A case study in the Pacific northwest. Energy Efficiency. 2020;**13**(6):1089-1104. DOI: 10.1007/s12053-020-09861-9

[47] Lupangu C, Bansal RC. A review of technical issues on the development of solar photovoltaic systems. Renewable and Sustainable Energy Reviews. 2017;**73**(February):950-965. DOI: 10.1016/j.rser.2017.02.003

[48] T. U. S. D. of E. and T. N. I. of B. Sciences, Homeowner's Guide to going solar. Available from: https://www.energy.gov/eere/solar/homeowners-guide-going-solar#:~:text=The size%2C shape%2C and slope, until it will need replacement.

[49] Atmaja TD. Facade and rooftop PV installation strategy for building integrated photo voltaic application. Energy Procedia. 2013;**32**:105-114. DOI: 10.1016/j.egypro.2013.05.014

[50] Qafleshi M. Solar photovoltaic roofs and Fasades of residential buildings as a compromise between design, cost and environment towards the sustainable city and community and climate action goals. In: UBT International Conference. UBT, Kosovo: University for Business and Technology; 2022. DOI: 10.33107/ubt-ic/2022.360

[51] Freitas S, Brito MC. Solar façades for future cities. Renewable Energy Focus. 2019;**31**:73-79. DOI: 10.1016/j.ref.2019.09.002

[52] Chen T, An Y, Heng CK. A review of building-integrated photovoltaics in Singapore: Status, barriers, and prospects. Sustainability (Switzerland). 2022;**14**(16) MDPI:10160. DOI: 10.3390/su141610160

[53] Asdrubali F, Ballarini I, Corrado V, Evangelisti L, Grazieschi G, Guattari C. Energy and environmental payback times for an NZEB retrofit. Building and Environment. 2019;**147**:461-472. DOI: 10.1016/j.buildenv.2018.10.047

[54] Moran P, O'Connell J, Goggins J. Sustainable energy efficiency retrofits as residencial buildings move towards nearly zero energy building (NZEB) standards. Energy and Buildings. 2020;**211**:15. DOI: 10.1016/j.enbuild.2020.109816

[55] Salem R, Bahadori-Jahromi A, Mylona A, Godfrey P, Cook D. Life-cycle

cost analysis of retrofit scenarios for a UK residential dwelling. Proceedings of the Institution of Civil Engineers: Engineering Sustainability. 2020;**173**(2):57-72. DOI: 10.1680/jensu.18.00055

[56] Bahadır Ü, Thomollari X, Kalfa SM, Toğan V. Energy and economic performance analysis of different ventilated wall construction types in residential buildings. Journal of Construction Engineering, Management & Innovation. 2020;**3**(4):305-318. DOI: 10.31462/jcemi.2020.04305318

[57] Sun X, Gou Z, Lau SSY. Cost-effectiveness of active and passive design strategies for existing building retrofits in tropical climate: Case study of a zero energy building. Journal of Cleaner Production. 2018;**183**:35-45. DOI: 10.1016/j.jclepro.2018.02.137

[58] Ma Z, Cooper P, Daly D, Ledo L. Existing building retrofits: Methodology and state-of-the-art. Energy and Buildings. 2012;**55**:889-902. DOI: 10.1016/j.enbuild.2012.08.018

[59] Balasbaneh AT, Yeoh D, Ramli MZ, Valdi MHT. Different alternative retrofit to improving the sustainability of building in tropical climate: Multi-criteria decision-making. Environmental Science and Pollution Research. 2022;**29**(27):41669-41683. DOI: 10.1007/S11356-022-18647-8/TABLES/7

[60] Rattanongphisat W, Rordprapat W. Strategy for energy efficient buildings in tropical climate. Energy Procedia. 2014;**52**:10-17. DOI: 10.1016/j.egypro.2014.07.049

[61] Lim T, Yim WS, Kim DD. Analysis of the thermal and cooling energy performance of the perimeter zones in an office building. Buildings. 2022;**12**(2):141. DOI: 10.3390/BUILDINGS12020141

[62] Bekdaş G, Aydın Y, Isıkdağ Ü, Sadeghifam AN, Kim S, Geem ZW. Prediction of cooling load of tropical buildings with machine learning. Sustainability. 2023;**15**(11):9061. DOI: 10.3390/SU15119061

[63] P. SNE. Secretaría Nacional de Energía de Panamá. 2020 [Online]. Available from: http://www.energia.gob.pa/mercado-energetico/?tag=79#documents-list.

[64] Plan Energético Nacional de Panamá. p. 343 [Online]. 2016. Available from: http://www.energia.gob.pa/energia/wp-content/uploads/sites/2/2017/06/Plan-Energetico-Nacional-2015-2050-1.pdf

[65] Gaceta Oficial. Resolución de la JTIA N°035. Panama: Gaceta Oficial; 2019. p. 61

[66] Gaceta Oficial. Resolución N°3142. Panama: Gaceta Oficial; 2016. p. 66

[67] CONUEE. Informe Nacional de Monitoreo de la Eficiencia Energética de Panamá. 2018. Nac. Chile: Unidas; 2020. p. 200. Available from: https://www.cepal.org/es/publicaciones/43612-informe-nacional-monitoreo-la-eficiencia-energetica-mexico-2018

[68] G. Oficial, Por Medio de LA CUAL SE APRUEBA el REGLAMENTO de EDIFICACIÓN SOSTENIBLE Para LA REPÚBLICA de PANAMÁ. V. 2. 2022, no. 29726, pp. 1-71. Panama: G. Oficial. Available from: https://www.gacetaoficial.gob.pa/pdfTemp/29726/96966.pdf

[69] Panama Green Building Council. Eco-Protocolo. Panama: Panama Green Building Council; 2017

[70] de Gabinete C. Resolución de Gabinete N° 1 de 7 de enero de 2020, Que aprueba el Plan Estratégico Nac. para el Desarro. la Ciencia, la Tecnol. y la

Innovación 2019-2024. Panama: Consejo de Gabinete; no. 28936; 2020. Available from: http://gacetas.procuraduria-admon.gob.pa/28936-B_2020.pdf

[71] Milligan B, Mehra M. Environmental Law Making and Oversight for Sustainable Development a Guide for Legislators. Law Division, Kenya: United Nations Environment Programme (UNEP); 2018

[72] Garmendia M, Grethy G, Autónoma U, León N, México N. Desarrollo sostenible y políticas públicas: enfoque de la ONU y ecología política Sustainable development and public policies: un approach and political ecology. Period. Semest. 2020;**6**:2020

[73] Bogdanovskaia I. The legislative bodies in the law-making process. Legislation Bodies Law-Making Process Introduction. 1999;**20**(2017):97-99

[74] United Nations. 17 Sustainable Development Goals. USA: United Nations; 2015. Available from: https://sdgs.un.org/goals

[75] The National Secretary for Energy. Law N° 43 of April 25th. Panama: The National Secretary for Energy; 2011

[76] Secretaría Nacional de Energía. Guía de construcción sostenible para el ahorro de energía en edificaciones. Panama: Secretaría Nacional de Energía; 2016

[77] Ministerio de Obras Públicas & Junta Técnica de Ingeniería y Arquitectura. Resolución JTIA N° 02: Reglamento de Edificación Sostenible. Panama: Ministerio de Obras Públicas & Junta Técnica de Ingeniería y Arquitectura; 2023

[78] De León L, Mora D, Carpino C, Arcuri N, Chen Austin M. A reference framework for zero energy districts in Panama based on energy performance simulations and bioclimatic design methodology. Buildings. 2023;**13**(2):315. DOI: 10.3390/BUILDINGS13020315

[79] Austin MC, Carpino C, Mora D, Arcuri N. A methodology to identify appropriate refurbishment strategies towards zero energy buildings in a hot and humid climate. Journal of Physics Conference Series. 2022;**2385**(1):012020. DOI: 10.1088/1742-6596/2385/1/012020

[80] Chacón L, Austin MC, Castaño C. A multiobjective optimization approach for retrofitting decision-making towards achieving net-zero energy districts: A numerical case study in a tropical climate. Smart Cities. 2022;**5**(2):405-432. DOI: 10.3390/smartcities5020023

[81] LEED. LEED Reference Guide for Building Design and Construction. USA: U.S. Green Building Council; 2019

[82] U.S. Green Building Council. Trademark Policy and Branding Guidelines. USA: U.S. Green Building Council; 2023. Available from: https://support.usgbc.org/hc/en-us/articles/11941118063891-USGBC-Trademark-Policy-and-Branding-Guidelines

[83] USGBC. LEED Zero Program Guide. Vol. no. May. USA: USGBC; 2020. p. 12

[84] Energy Star. Source Energy: Technical Reference, no. August. USA: Energy Star; 2013. pp. 1-19. Available from: https://portfoliomanager.energystar.gov/pdf/reference/Source Energy.pdf?ccac-8037

[85] IFC. EDGE User Guide, version 3.0.a. USA: Edge; 2021. Available from: https://edgebuildings.com/learning/user-guides/

[86] EDGE. EDGE Methodology Report. USA: EDGE; 2022

[87] "EDGE Software App." Available from: https://edgebuildings.com/?lang=es.

[88] Austin MC, Chung-Camargo K, Mora D. Review of zero energy building concept-definition and developments in latin america: A framework definition for application in Panama. Energies. 2021;**14**(18):1-29. DOI: 10.3390/en14185647

[89] González J, Austin MC, Mora D. Estimation of Residential Energy Consumption Accounting for Occupant-Behavior and Building Characteristics: A Coupled Statistical Approach with Reduced Survey and Simulation Tool. Vol. 347 LNNS. Lecture Notes in Networks and Systems; 2022. pp. 164-171. DOI: 10.1007/978-3-030-90033-5_18/COVER

Chapter 5

# Ocean Energy Harvesting History and Technologies

*Sijing Guo, Shuo Chen and Mudasar Zahoor*


## Abstract

Marine energy is abundant and has been around since the very beginning. The energy density is much higher than those of solar energy and wind but we have yet to commercialize the product. Three main challenges are proposed by many peer-reviewed journal papers: Low reliability, Levelized Cost of Energy (LCoE), and energy production. I would like to get deep into the source of the issues in the harvester development, discuss the potential improvements, and application of those converters, and lastly focus on the commercialization along with the renewable energy policies. This will help the audience understand how it all started, where we are currently, and how far we are from commercialization.


**Keywords:** energy harvesting, levelized cost of energy, marine energy, sustainability, blue economy

## 1. Introduction

### 1.1 Overview

The Earth's surface is covered by the ocean for more than 70% and has abundant wave, tidal, and thermal energy sources. However, ocean energy harvesting was initially considered too costly compared to other types of renewable energy, such as wind and solar energy. Recently, the dependency on fossil fuels and global warming have exerted greater pressure, and increasing attention has focused on the vast energy resources in the ocean, especially in coastal regions of the United States. This chapter demonstrates an overview of the wave, tidal, ocean current, and ocean thermal energy conversion energy and presents potential future innovations. The concepts of each energy, the energy potential, and the development history of various ocean energy sources are elaborated.

### 1.2 Tidal energy harvesting

Tidal energy harvesting is a gift from nature. It comes from the Moon's gravitational pull, which then leads to periodic ocean tides [1]. The water goes up and down inside and outside the bay, rivers, and harbor area [2]. The river necking area with shallow water where water can travel fast is one of the most efficient ways to harvest energy from the tides.

Due to the unknown underwater geography, estimation of the global tidal stream energy potential has always been a challenge. In 1993, it was estimated to be 3 TW (terawatt), with 1 TW found in shallow waters [2]. The full potential of stream energy can peak while the current is traveling between 2 and 2.5 m/s, which is usually found along the coastlines worldwide [3]. In Europe, there have been 106 sites identified with great potential for tidal energy harvesting with an estimated output of 48 TWh (terawatt-hour) annually with full utilization. Similar sites were also found near the British Isles and the English Channel, and this has shown great potential and economic benefit to continue this research and commercialization [4].

Harnessing tidal energy is not a new idea. Tidal forces were used by people in early 900 A.D. in Britain and France [5], where tidal-powered mills were built by placing a barrier at the entrance of a basin. The basin would be filled with the incoming tide, and when the tide went out, the trapped water was released to drive waterwheels or paddlewheels, which generated mechanical power, and further electricity. The main purpose of these mills was to grind grains and corn. The major development in tidal energy happened in the twentieth century. Back then, the La Rance Tidal Power Station was constructed in France and started operating in 1966 [6]. This station, located on the Rance River estuary, has a tidal range of up to 13.5 meters during equinox tides and uses a 740-meter-long barrage with 24 two-way 10 MW turbines, generating a total of 240 MW. It can supply power to about 4% of homes in Brittany and remains the world's largest operational tidal plant, receiving much attention as an innovative project in tidal energy. Other operated facilities include Sihwa Lake Tidal Power Station in South Korea [7] and Tidal Lagoon Swansea Bay in the UK [8], which were great examples showing promising growth of tidal energy development and should also be suitable near the coastlines close to the UK, Canada, and part of Asia.

Tidal energy is still a promising but underestimated resource. With the ongoing research and improvements in tidal stream turbine technology, many coastal countries have started to better understand and apply tidal energy as an alternative energy source.

### 1.3 Wave energy harvesting

Wave energy is proving to be the most commercially advanced of the ocean energy technologies [9]. Various innovative devices were developed to capture and collect the kinetic and potential energy from the wave movements. Corresponding technologies include oscillating water columns and attenuators. The former uses the rise and fall of water to drive turbines, and point absorbers, converting the vertical motion of waves into mechanical energy. The latter captures the motion of waves along their length to produce energy.

**Figure 1** shows wave energy levels in kilowatts per meter (kW/m) along coastlines around the world. The measurement is based on the distance from one wave peak to the next. The figure shows that wave energy potential changes a lot by latitude, with the highest levels seen in areas above 40° latitude in both hemispheres. These areas are good candidates for wave energy development. For example, in the U.S., the wave energy in Alaska reaches up to 67 kW/m, while the medium level of wave energy in Hawaii is only 10–15 kW/m, and the Gulf Coast has almost none. Other high-energy areas exist in Japan's coastal waters (43 kW/m), parts of Europe (34 kW/m), and southern Australia (73 kW/m). This spread shows that places like Alaska and southern Australia are especially suitable for wave energy projects.

*DOI: http://dx.doi.org/10.5772/intechopen.1008180*

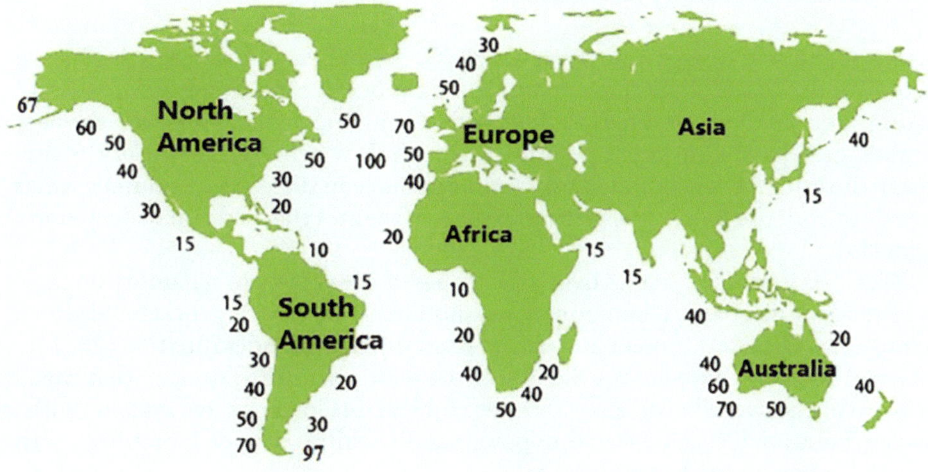


**Figure 1.**
*Wave energy levels in kW/m crest length (reprinted (adapted) with permission from Ref. [9] Copyright (2009)).*

Numerous methods have been developed to convert wave energy into usable electricity [10–12]. Early developments such as the Salter Cam wave energy converter in the U.S. [13] and a floating hinged system in the UK [14] have both shown merits in the deployment. Other advanced developments including the underwater generator by Kayser [15, 16], a wave-powered pump [17], and a pneumatic wave converter are all great examples of the early development of wave energy harvesters. The first four systems are meant for large-scale energy production, while the last one is used as a low-power supply for navigation buoys.

As the technologies are getting more matured over the years, under optimal conditions, 90% of the wave energy can be utilized and converted to electricity [18]. Currently, only 35% of the conversion efficiency is a suitable goal for product development including all auxiliary processes. The challenge remains to be the variable and unpredictable power source to the power generation system hardware and the transmission to the grid. The capital investment to build such infrastructure is posing challenges and therefore resulting in a slow commercialization speed.

From the research aspect, three stages usually govern the path from research and development to commercialization. The first stage usually represents a small laboratory prototype with full functionality but less efficiency. When the capital starts flowing in from the sponsors, the second stage starts. Government funding or private equities are usually helpful, and the extra resources enable the researchers to build on a larger scale and at the same time be able to demonstrate its capability and show promising results. Lastly, in the final stage, the prototype turns into an actual product, starts the production launching process, and becomes grid-connected. It usually goes through the engineering phase process and is peer-reviewed by professionals in the industry for design validations for public safety and tests for its full functionality. However, up to today, only about 20 prototypes have reached the advanced testing phase and even fewer are close to commercialization. However, with the recognition of importance and support from industry partners and funding agencies, marine energy is expected to grow tremendously in the following years.

### 1.4 Ocean thermal energy harvesting

Ocean Thermal Energy Conversion (OTEC) stores heat in the ocean during the daytime under the sunlight. Due to the size of the ocean coverage, it automatically makes the ocean the largest natural heat collector on the planet. However, the deeper water tends to have a lower temperature due to the low penetration rate of the sunlight in the water. A low boiling point fluid can move in the vertical column, which extends to both the hotter zone closer to the surface and the colder area deeper in the ocean [19].

OTEC has great advantages over other types of converters in reliability due to the absence of mechanical components in motion. Therefore, it generates almost no impact on the environment and can be used for saltwater desalination [20, 21]. However, due to the low-energy density of the temperature difference, we normally see 2–7% of the overall efficiency and therefore results in expensive systems with less power generation [20, 22, 23]. Other potential difficulties include biofouling, corrosion, and extreme weather effects [24].

The OTEC idea was first proposed in 1881 by the French scientist Jacques-Arsène d'Arsonval, who proposed that the temperature difference could be used to generate electricity. In 1930, his student George Claude was finally able to build the first OTEC plant in Cuba, which could produce 22 kW of power [25]. Due to the oil crisis where countries started to pay more attention to marine energy harvesters, the first experimental OTEC plant was built in Hawaii in 1974 [26]. India and Japan also built a 1 MW OTEC system offshore near Tuticorin [27] and Okinawa, respectively [28]. As the technologies become more promising, the current commercialized systems are aiming for 50 MW, and the goal is to offset the production cost.

### 1.5 Summary

Three primary ocean energy harvesting technologies, such as tidal energy, wave energy, and ocean thermal energy, were introduced in this chapter. The concepts, energy potential, and development history of these technologies across various countries were covered. In the three technologies, tidal energy harvesting is the most mature technology that several commercial generating plants were established with reliable technologies. Wave energy harvesting is still in the demonstration stage. Several critical challenges are yet to be conquered before its extensive utilization. OTEC is currently in the research and experimental stage. The appropriate area mainly focuses on tropical and subtropical. Different from the above two technologies, OTEC holds the potential to provide byproducts such as fresh water. However, the high costs and complex technological requirements have so far prevented large-scale commercial deployment.

## 2. The tidal/wave/current energy evolutions

### 2.1 Overview of the technologies

Marine tidal, wave, and current energy are some of the most abundant energy resources in the world. The wave energy alone could reach 2640 terawatt-hours just near the U.S. shoreline, which is equivalent to 63% of the total utility-scale power generation in the U.S. [29]. The massive amount of energy that is stored has yet to be exploited and a fraction of the energy harvester could have a profound impact on the population near the coastline. Therefore, this chapter focuses on a few technologies developed in the past years, and the innovations are crucial to the growth of the marine energy harvesting industry.

Dr. Falcao, one of the most successful marine energy scientists, categorizes the wave energy converters (WECs) based on their forms and working principles to devices, such as oscillating water columns, overtopping devices, and oscillating body devices [30].

An oscillating water column utilizes the change in pressure inside a column to drive the turbine for power generation. The overtopping device then stores the water in a reservoir at the wave crest and slowly releases the water through a turbine system, which pushes for the turbine rotation and generates power [31]. This technology, widely used in countries like Scotland and Japan, has proven effective for large-scale energy conversion [32] and is normally capable of generating power from 500 kW and 1 MW [33].

Overtopping devices convert energy from the overtopping water, collected in a reservoir, which then flows down and then rotates the turbine for energy production. The Wave Dragon, for instance, is a great example of an overtopping device developed in Denmark [34]. It was then connected to the grid in 2004 and showed promising results.

Finally, the oscillating body devices could also be one of the most common harvesters including a point absorber, attenuators, pitching devices, etc. In general, oscillating body devices capture all power take-off (PTO) system that uses reciprocating waves as a power source, which empowers an oscillating motion then to generate electricity [35]. Attenuator devices, for example, were developed to convert energy from its continuous bending motion, which is powering the hydraulic pumps inside. The Pelamis attenuator wave energy converter device [36] was developed in the UK and is capable of generating 5 MW and has been deployed in Portugal in 2008 [37]. On the other hand, the point absorber is also popular in real-life testing and has shown promising results. The AWS Ocean Energy company has developed a point absorber in the Netherlands [38]. It is a submerged device that responds to underwater pressure changes from passing waves and is anchored to the seabed. The wave's vertical oscillation contributes to the power generation directly. The AWS point absorber was tested in Portugal and was able to generate 250 kW with scalability [39].

### 2.2 LCOE and standard reference models

Research and development based on the above energy harvesters motivated the industry to the commercialization of marine tidal, wave, and current energy. Therefore, the vast majority of device development eventually comes down to the unit cost, or the Levelized Cost of Energy (LCOE). LCOE has been widely adopted in the renewable energy industry describing the average cost of producing power over the converter's lifetime subtracting the operation maintenance cost. Therefore, the developer should innovate and pay attention to the sustainability of the developed idea to beat the cost target.

To better serve the marine energy community, the U.S. Department of Energy (DOE) partnered with The National Laboratories to develop open-source marine hydrokinetic (MHK) point designs and named them reference models (RMs). The standards not only listed out the most popular tidal, wave, and current energy harvester styles but also provided a cost reference for future development (**Figure 2**).

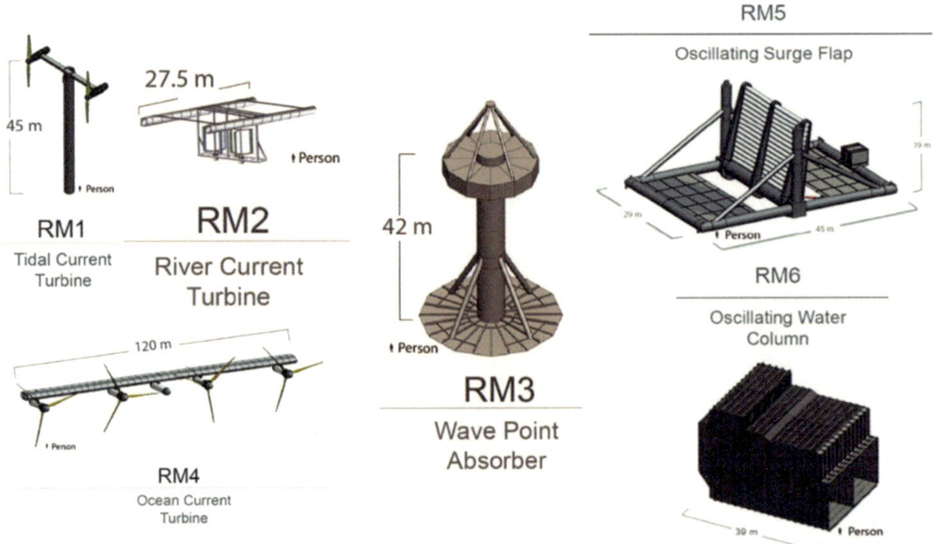


**Figure 2.**
*The new reference models (RMs) published by the Department of Energy target standardizing product development cost targets. (Reprinted (adapted) with permission from Ref. [40]).*

The RM1, RM2, and RM4 are harvesters based on water current and therefore utilize a water turbine for energy harvesting. These methods are generally effective and have shown promising results at dams and narrow creeks where faster water current exists. The energy harvesting method consists of transferring kinetic energy in the water flow to the water turbine for maximum energy output. The latest research also includes investigating the funneling effect on the turbine in the north and south poles to power low-energy devices and is soon to be published. The benefit of the water turbine is the reliability and cost-effectiveness of a simpler structure and wide deployment zone from the various designs of the turbine blades.

The RM5 is a flap device or surge converter. A hinged single-flap design example includes a commercialized Aquamarine Power Oyster 1 and 800, which harvest energy from driving the PTO near the flap hinge through the wave oscillation [41]. The hinged surge device could effectively harvest energy from the wave due to its simple and stable structure. The other main research path on the surge device contains a double-flap floating surge that harvests energy from the double-flap's relative motion [42]. The benefit of the floating design includes but is not limited to a higher energy harvesting rate, and work with all depths since it uses mooring lines rather than anchors.

RM6 is an Oscillating Water Column (OWC) concept that started in the 1940s and the first paper was published in 1978 [43]. Most OWCs utilize the tidal or wave rise and fall to create the air pressure difference between an enclosed chamber and the external environment. The air goes through a funneled turbine system, harvests kinetic energy, and converts it to electricity. The OWC has a unique feature of easy scaling, and the power produced is proportional to the size of the harvester.

Lastly, the RM3 is a point absorber style energy harvester and harvests energy from the oscillating wave motion. The point absorber usually consists of two styles, a single- or a two-body system. A single-body point absorber usually mounts directly to the seabed, with the buoy floating on the water surface. The oscillating wave energy is then transferred to the buoy, which is then strung to the PTO near the ocean bed. The wave's kinetic energy is directly driving the motor and generating electricity [44]. The two-body point absorber is a much more common style of energy converter that has been heavily investigated by scientists over the years. The first body floats on the water surface and the second body, with a large mass, is submerged in the water and behaves like a stator. As the waves arrive at the system, the buoy is excited by the wave and induces a vertical oscillation. The second body, however, remains relatively stationary as the buoy is driving the PTO's generator through a lead screw or similar mechanisms that convert linear motion to rotation that can be harvested by the generator [35]. The generator hence rotates in both directions as the buoy oscillates vertically.

Since the point absorber motion relies heavily on the wave input without systematic internal damping, reliability can be an issue and was analyzed through Finite Element Analysis (FEA) modeling to help understand the fatigue failures [45]. Two typical methods are used to reduce the fatigue of the PTO. One is through a Mechanical Motion Rectifier (MMR), which is a specialized gearbox designed to unify the bidirectional rotation at the end of a lead screw through the combinations of clutches. This method relies only on a gearbox to achieve the rotation direction unification and can be robust and cost-effective (**Figure 3**) [46].

The second method is to add a rotor in the system to maintain the high-speed rotation such as a flywheel. This can preserve the rotation during the low-energy state of the ocean wave when it oscillates across the zero-amplitude zone. However, the flywheel design relies on the success of the MMR gearbox since the rotation has to be unified before the flywheel can maintain its speed. This can not only boost energy generation but also prevent the system from experiencing high fatigue stress from the constant oscillating power output (**Figure 4**) [47].

Lastly, a controllable gearbox can be used to pause the system operations during severe weather conditions based on an Active-MMR system [48]. This design utilizes electromagnetic clutch systems to recreate the MMR system but with controllable clutches. It can disconnect the power generation side (output side), from the wave input side, without allowing the energy to flow through the system (**Figure 5**).

The existing RMs set a standard in marine tidal and wave energy product development on LCOE. With the baseline product provided, the innovation could follow a

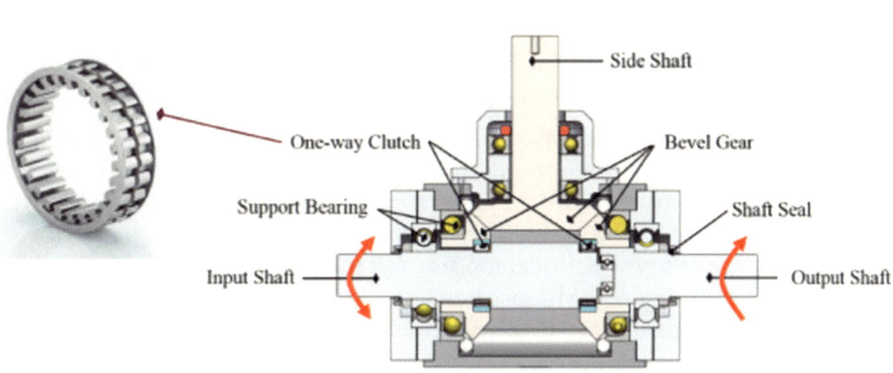


**Figure 3.**
*The above shows a figure of a typical MMR design. The input shaft rotation is bidirectional and with the implementation of the one-way clutches, the output shaft only rotates one direction. (Reprinted (adapted) with permission from Ref. [46]).*

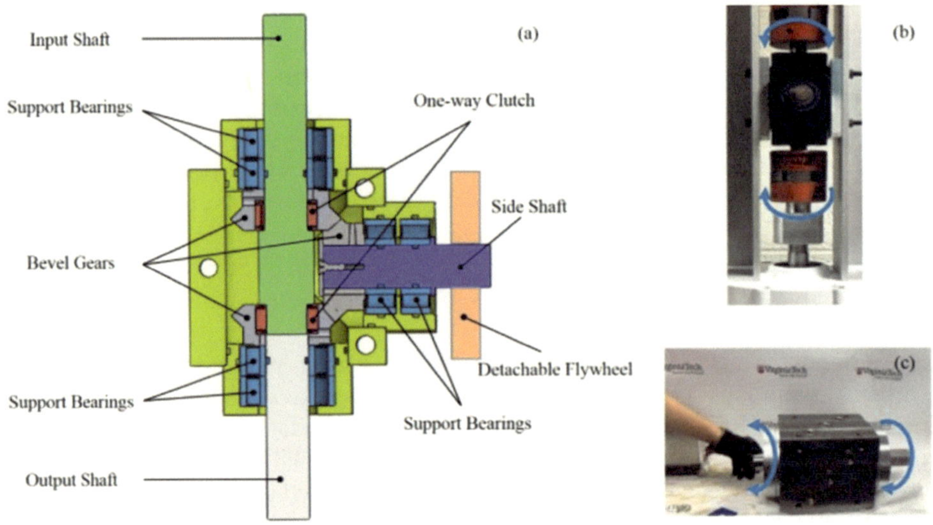


**Figure 4.**
*MMR-based Flywheel design to improve power output and reliability. (Reprinted (adapted) with permission from Ref. [47]).*

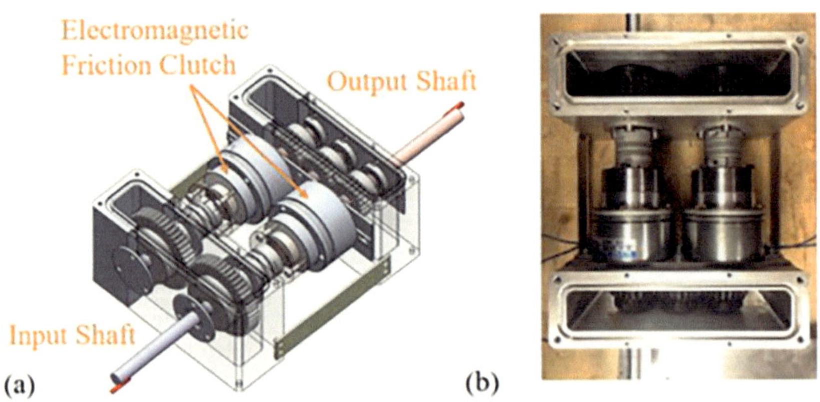


**Figure 5.**
*Active mechanical motion rectifier. (Reprinted (adapted) with permission from Ref. [48]).*

guided path and become more applicable to real life. One of the latest innovations based on the RMs is a "Hybrid System." Researchers worked on the combination of a marine turbine and a point absorber system to reduce the LCOE and improve system reliability. The concept is simple, but the engineering side is complex due to the nature of the turbine's unidirectional motion and the point absorber's bidirectional rotation. However, since the MMR gearbox has been developed, the turbine can naturally fit in the scope due to MMR's rotation unification feature (**Figure 6**) [49].

This design incorporates the reliability of the turbine and the simple design of an MMR gearbox. Both the turbine and the point absorber use only one PTO and therefore save on the production and maintenance costs by a significant amount. Furthermore, the system's reliability is improved by the high and constant power output from the system. This is achieved through the implementation of the turbine,

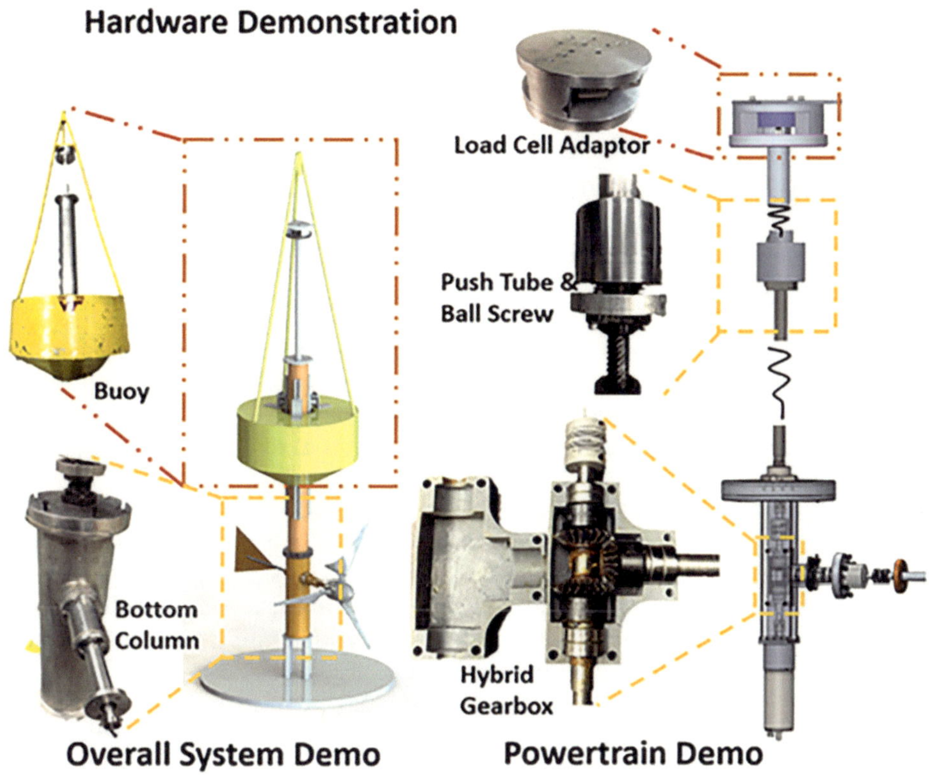


**Figure 6.**
*The hybrid marine energy converter that harvests ocean currents and waves simultaneously. (Reprinted (adapted) with permission from Ref. [49]).*

which increases the general power output level from zero watt as the point absorber goes through the zero-amplitude phase. Therefore, the Peak-to-Average (PAR) ratio decreases as the turbine in the hybrid converter becomes operational, which is indicative of improved system reliability.

This concludes the latest designs and innovations on the reliability improvements regarding the point absorber style wave energy harvesters utilizing a gearbox. The commercialization is dependent on the reliability of the LCOE calculation. The DOE's RMs provide a good starting point for all scholars, and they can be implemented by most wave energy harvesting researchers.

### 2.3 Summary

Tidal, wave, and current energy are considered one of the most abundant renewable and sustainable sources of power in the ocean, and they can produce a significant portion of the electricity demand. The technologies are evolving from continuous research on the efficiency, LCOE, and minimizing the environmental impacts. However, it is also experiencing numerous challenges. For instance, low reliability from harsh environments, maintenance issues, and power grid integration. Still, with all factors considered, ocean energy still represents a promising avenue for the expansion of renewable energy and reducing the dependence on fossil fuels.

## 3. The future of ocean energy

Sustainable energy solutions have always been a strategic resource for all countries. It is crucial to secure a level of renewable energy sources and maintain a high level of development constantly.

### 3.1 Wave energy converters (WECs)

WECs, as mentioned, convert the kinetic and potential energy to electricity through a PTO system.

- Oscillating Water Columns (OWCs): The first full-scale OCW plant was built by an Italian company Waveenergy.it Ltd. at the Port of Civitavecchia with an installed capacity of 2.2–2.5 MW [50]. This system is relatively robust and can be easily scaled up, and therefore has a massive potential to be commercialized.

- Point Absorbers: The PB3 PowerBuoy® by Ocean Power Technologies is a notable example, capable of providing continuous power in remote offshore locations [51]. The example represents the possibility of a larger scale point absorber system and research has identified the challenges on the reliability side. However, the point absorber is one of the higher-efficiency wave energy converters on the market and its potential shall not be ignored.

- Attenuators: The attenuators are relatively simple and more reliable. Pelamis Wave Power pioneered this technology, though the company has since ceased operations [36, 52].

### 3.2 Ocean thermal energy conversion (OTEC)

OTEC utilizes the temperature differences and demonstrates its reliability.

- Closed-cycle OTEC: An enclosed chamber promoting heat transfer internally. The Makai Ocean Engineering OTEC plant in Hawaii demonstrates this technology at a small scale of approximately 100 KW [53].

- Open-cycle OTEC: It utilizes the warm seawater itself as the working fluid. While less common, this method has the added benefit of producing desalinated water [54].

These technologies show their promising prototype results and high reliability. However, maintenance and cost-effectiveness can be challenging.

### 3.3 Salinity gradient energy/power

Salinity gradient energy (SGE) harvests energy from different salt concentrations between two fluids. For instance, SGEs can harvest the difference in salt concentration between freshwater and seawater to generate electricity; therefore, they were usually placed in between rivers, lakes, and oceans. Two main approaches are being developed:

- Pressure Retarded Osmosis (PRO): This approach converts the osmotic pressure of saline water to hydraulic pressure, which can then be used to drive a turbine and therefore, generate electricity [55]. The world's first osmotic plant with a capacity of 10 kW was opened by Statkraft, a state-owned hydropower company, on November 24, 2009 in Tofte, Norway [56].

- Reverse electrodialysis (RED): The method uses ion-selective membranes to separate salt ions, creating an electric current. The REDStack pilot plant in the Netherlands demonstrates this technology's potential [57].

### 3.4 Offshore solar installations

The idea of floating solar farms is an innovative way to utilize ocean space for renewable energy production:

- Floating Photovoltaic (FPV) Arrays: Large-scale floating solar installations are already operational in still/calm waters. The largest operational FPV project (320 MW), Dingzhuang solar farm, is located on a reservoir in eastern China. The Sunseap Group also is planning to develop a 5 GW solar farm off the coast of Indonesia [58].

- Solar-Wind Hybrids: It is an interesting idea that combines offshore wind turbines with floating solar panels to maximize energy production with existing technologies. Companies such as Oceans of Energy, a Dutch company, are planning to construct the first commercial offshore solar-wind hybrid project in the North Sea [59].

### 3.5 Marine biomass energy

The concept refers to the potential to generate energy from vegetal biomass found in the ocean. The biomass energy resource for electricity production is a growing field with significant potential:

- Algal biofuels: Large-scale growing of algae for biofuel production can be an alternative to fossil fuels rather than directly converting to electricity. A biotech company named Algenol is developing an efficient technology, to convert Algal to fuel for transportation [60].

- Biogas from seaweed: Seaweed can generate Biogas, and the conversion is completed through the anaerobic digestion process. Seaweed is rich in carbon dioxide and methane elements that can be immediately used for electricity generation or as a fuel source to power cars. The European Union (EU)-funded MacroFuels project has made progress in the Biogas sector [61].

## 4. Use of AI and machine learning in ocean energy

Artificial Intelligence (AI) and Machine Learning (ML) technologies are in fact playing a critical role in advancing ocean energy technologies sooner than we thought. These techniques with proven advantages in other fields can be applied from design and optimization to operation and maintenance.

### 4.1 Design optimization

The design of wave energy converters (WECs) and tidal turbines utilizes AI technologies for shape optimizations and cost analysis. For instance, a neural network algorithm can quickly evaluate thousands of design iterations, reduce the material as it goes forward, and converge to the most efficient shapes and configurations for specific ocean conditions [62].

### 4.2 Predictive maintenance

ML models are being researched to predict equipment failures, reducing system downtime and maintenance costs. By analyzing data from sensors in the ocean energy devices, the ML models can predict the failure mode, analyze the failure characteristics, and predict the failure time.

### 4.3 Resource assessment and forecasting

AI/ML-powered systems can be applied to improve the accuracy of ocean energy resource assessments. The algorithms can process vast amounts of oceanographic data to predict wave heights, tidal currents, and other relevant parameters with greater precision than traditional methods [63], therefore, paving the way for highly optimized and cost-effective designs.

### 4.4 Grid integration and energy management

As ocean energy systems become more prevalent, AI is playing a crucial role in managing their integration into power grids. ML algorithms can optimize power output based on grid demand and market prices, enhancing the economic viability of ocean energy projects [64].

### 4.5 Environmental monitoring

AI-based image and sound recognition systems are currently being developed to evaluate human activities' impact on marine lives. These systems can automatically detect and track marine mammals, helping to protect regional wildlife existence [65].

Even though AI and ML use in ocean energy is still in its infancy, it has already started showing promising results. Despite the slower development of marine energy development, the new technologies could make marine energy harvesting more competitive with other renewable energy sources.

However, even the greatest technologies come with downsides:

a. Environmental impact: It is important to ensure that these technologies do not adversely affect marine ecosystems. There is ongoing research that focuses on minimizing impacts on marine life and habitats [66].

b. Cost competitiveness: In order to bring ocean energy technologies to the commercial platform, it is imperative that along with improved efficiency, the costs should come down as well. Many ocean energy technologies are still more expensive than conventional renewables like solar and wind. Continued research and development are needed to reduce costs and improve efficiency [67].

c. Grid integration: Developing infrastructure to transmit electricity from offshore installations to onshore grids remains a significant challenge, particularly for remote deep-water installations [68].

d. Durability: Ocean environments are harsh and corrosive. This shows the better materials usage and practical designs can withstand these extreme weather conditions.

Despite the challenges, the future of ocean energy still looks promising. With the continuously growing technology and better system-level optimizations, the marine energy industry continues to grow. The International Energy Agency estimates that ocean energy development will top 63 GW in 2050, contributing significantly to global renewable energy production [69].

The development of ocean energy technologies is not just sustainable electricity production, but also creates jobs, boosts the economy, and promotes technological innovation. As climate change is becoming more severe over the years, marine energy harvesting may well prove to be a crucial part of the solution.

## 5. Powering the blue economy

The "Blue Economy" has become a popular topic in the past few years. It focuses on the commercialization of the technologies and how to convert concepts into real business ideas.

In this regard, we sift through the relevant applications as follows:

### 5.1 Offshore marine aquaculture

Offshore marine aquaculture shows a sustainable idea to farm seafood with marine energy. Therefore, it seems logical that ocean energy generation is used to power these operations. Wave energy converters or floating wind turbines can provide electricity for feeding systems, monitoring equipment, and even processing facilities [70].

### 5.2 Desalination

Desalination plants can be powered by ocean energy, thereby proving to be a sustainable source of potable water. Ocean Thermal Energy Conversion (OTEC) systems are particularly suitable for this application because they can produce both electricity and desalinated water as a byproduct [71].

### 5.3 Coastal resiliency

The coastal communities face increasing challenges due to more frequent extreme weather events including rising sea levels. Wave energy converters can be integrated into breakwater structures, serving the dual purpose of energy generation and coastal protection [72].

### 5.4 Marine research and exploration

Understanding the marine environment is crucial for its sustainable management. Ocean energy can power remote research stations, subsea observation systems, and underwater vehicles. This will provide a suitable platform to collect long-term data in somewhat inaccessible areas [73].

### 5.5 Sustainable island communities

Providing power to island nations and remote coastal communities is challenging. Ocean energy generation systems offer a promising solution, providing a renewable and local power source. Yates et al. [74] have discussed the integration of ocean energy including offshore solar and other storage systems to create resilient microgrids for island communities.

### 5.6 Seabed mining

Seabed mining will play a role in the future of the Blue Economy. Ocean power can be used for offshore mining operations, potentially reducing their environmental effect compared to diesel-powered systems. However, careful environmental assessment and regulations are crucial to ensure sustainable practices.

### 5.7 Ocean cleanup technologies

Marine pollution has increased manifold, especially with the increase in global trade and shipping. Powering autonomous collection vessels as well as processing platforms using ocean energy seems plausible. Moreover, the ocean cleanup project is exploring using ocean currents not just for collecting pollutants like plastics but also as a potential energy source for its systems [75].

Integrating ocean energy into the Blue Economy comes with significant challenges. There are material and technical issues related to device durability in harsh marine environments, grid integration, and energy storage. Additionally, potential conflicts with other ocean uses, such as fishing or shipping, require careful spatial planning.

Having said that, the opportunities are significant as well. By powering various sectors of the Blue Economy, ocean energy can contribute to multiple United Nations Sustainable Development Goals [76], including:

- Affordable and Clean Energy (SDG 7)
- Industry, Innovation, and Infrastructure (SDG 9)
- Life Below Water (SDG 14).

With the organic growth of technologies and decrease in costs, ocean energy has become a more realistic idea compared to a decade ago. These advancements came from the efforts of the worldwide scientific community and are not slowing down. With the implementation of AI and ML, the growth of technology has never been so rapid. These innovations in the marine energy industry are beneficial to the world's citizens and should be supported. Marine energy has become not only a strategic gadget that advances countries over others but also a benefit to every person on the planet. It is to all the authors' belief that the bloom of the marine energy industry is coming soon and all of us will benefit from this success.

## 6. Conclusion

The marine energy industry highlights rich innovation and resilience, reflecting the renewable energy researchers' forward quest for a sustainable future. From the early stage of the tidal mills that utilize the motion to today's WEC devices that convert to electricity, the growth is significant.

However, the challenges in the field still remain, including the environmental impacts, low LCOE, and regulatory guidelines. What's certain is that the collaborations between the government, industrial partners, and research institutions are pushing this industry forward to overcome the listed obstacles. Together, it would unlock the full potential of marine energy and make it a viable contributor to the global renewable energy market.

As the world is shifting towards renewable energy, the marine energy industry could be leading the way due to its abundance around the continents. It could be the turnkey solution to worldwide energy shortages.

## Acknowledgements

The authors would like to acknowledge the help of many researchers who provided access to the pictures in the chapter. They were a great help in composing this effort.

## Conflict of interest

No potential conflict of interest was reported by the authors.

## Author details

Sijing Guo[1*], Shuo Chen[2] and Mudasar Zahoor[3]

1 Wuhan University of Technology, Wuhan, China

2 Johns Hopkins University, Baltimore, USA

3 Array Technologies, Chandler, USA

*Address all correspondence to: sijingg@whut.edu.cn

## References

[1] Alberdi M, Amundarain M, Garrido AJ, Garrido I, Maseda FJ. Fault-ride-through capability of oscillating-water-column-based wave-power-generation plants equipped with doubly fed induction generator and airflow control. IEEE Transactions on Industrial Electronics. 2011;**58**(5):1501-1517

[2] Greaves D, Iglesias G. In: Wiley J, editor. Wave and Tidal Energy. Hoboken, New Jersey, USA: John Wiley & Sons Limited; 2018

[3] Ocean S. Ocean Energy: State of the Art. Brussels, Belgium: EU Project Report; 2012

[4] Mofor L, Goldsmith J, Jones F. Ocean energy: Technology readiness, patents, deployment status and outlook. In: International Renewable Energy Agency (IRENA) Report. Paris: IRENA; 2014

[5] Irwin DA. Against the Tide: An Intellectual History of Free Trade. Princeton, New Jersey, USA: Princeton University Press; 2020

[6] Sheth S, Shahidehpour M. Tidal energy in electric power systems. In: IEEE Power & Energy Society General Meeting. Piscataway, New Jersey, USA: IEEE; 2005

[7] Park ES, Lee TS. The rebirth and eco-friendly energy production of an artificial lake: A case study on the tidal power in South Korea. Energy Reports. 2021;7:4681-4696

[8] Petley S, Aggidis G. Swansea Bay tidal lagoon annual energy estimation. Ocean Engineering. 2016;**111**:348-357

[9] Burman K. Ocean Energy Technology Overview. Golden, Colorado, USA: National Renewable Energy Laboratory (NREL); 2009

[10] Kramer M. The Wave Energy Converter: Wave Star, a Multi Point Absorber System. Technical Report. Bremerhaven, Denmark: Aalborg University and Wave Star Energy; 2008

[11] Rynne TM. Ocean Wave Energy Conversion System. U.S. Patent 5,136,173. Washington, DC: Scientific Applications & Research Associates, Inc.; 1991

[12] Sheindlin AE, Kovbasyuk VI. Open-cycle magnetohydrodynamic energy conversion: State of the art and prospects for the future. Perspectives in Energy; (United Kingdom). 1991;**1**(2):33-60

[13] Salter SH. Wave power. Nature. 1974;**249**(5459):720-724

[14] Wolley M, Platts J. Energy on the crest of a wave. New Scientist (1971). 1975;**66**:241-243

[15] Kayser H. Energy generation from sea waves. In: Proceedings of the Ocean 1974 IEEE Conference. NS, Canada: Halifax; 1974. pp. 240-243

[16] Isaacs JD, Wick GL, Schmitt WR. Utilization of energy from ocean waves. In: Proceedings of the Wave and Salinity Gradient Energy Conversion Workshop. USA: University of Delaware; 1976. pp. 1-36

[17] Isaacs JB, Castel D, Wick GL. Utilization of the energy in ocean waves. Ocean Engineering. 1976;**3**:175-187

[18] Jones WJ, Ruane M. Alternative Electrical Energy Sources for Maine, Appendix I, Wave Energy Conversion by J. Mays. Report No. MIT-E1 77-010.

Cambridge, Massachusetts, USA: MIT Energy Laboratory; 1977

[19] Girgis MA, Siegel JM. Open-Cycle Ocean Thermal Energy Conversion. Florida Solar Energy Center, Technical Report, No: FSEC-FS-28-83. Cocoa, Florida, USA; 1983

[20] Rahman S. Alternate sources of energy. IEEE Potentials. 1988;7:22-25

[21] Pelc R, Fujita RM. Renewable energy from the ocean. Marine Policy. 2002;**26**:471-479

[22] Kajikawa T, Agawa T. System characteristics of low temperature difference power generation. IEEE Translation Journal on Magnetics in Japan. 1978;**98-B**:773-780

[23] Heydt GT. An assessment of ocean thermal energy conversion as an advanced electric generation methodology. Proceedings of the IEEE. 1993;**81**:409-418

[24] Panchal CB, Larsen-Basse J, Berger L, Berger J, Little B, Stevens HC, et al. OTEC Biofouling-Control and Corrosion-Protection Study at the Seacoast Test Facility: 1981-1983. Argonne, Illinois, USA: Argonne National Laboratory Report ANL-IOTEC-TM-5; 1985

[25] Takahashi MM, Kitazawa K, Snowden P. Deep Ocean Water as our Next Natural Resource. Tokyo, Japan: Terra Scientific Publishing Company; 1991

[26] Energy Information Administration Staff. Average Retail Price of Electricity to Ultimate Customers by End-Use Sector, by State, December 2007 and 2006. Washington, D.C., USA: Energy Information Administration; 2008

[27] Ravindran M, Abraham R. The Indian 1 MW demonstration OTEC plant and the development activities. In: Proceedings of the Ocean '02 MTS. Piscataway, New Jersey, USA: IEEE; 2002. pp. 1622-1628

[28] Cheng C, Blakers A, Stocks M, Lu B. 100% renewable energy in Japan. Energy Conversion and Management. 2022;**255**:115299

[29] U.S. Energy Information Administration - EIA - Independent Statistics and Analysis. Available from: https://www.eia.gov/energyexplained/hydropower/wave-power.php

[30] de Falcão AFO. Wave energy utilization: A review of the technologies. Renewable and Sustainable Energy Reviews. 2010;**14**(3):899-918

[31] Contestabile P, Crispino G, Di Lauro E, Ferrante V, Gisonni C, Vicinanza D. Overtopping breakwater for wave energy conversion: Review of state of art, recent advancements and what lies ahead. Renewable Energy. 2020;**147**:705-718

[32] Tseng RS, Wu RH, Huang CC. Model study of a shoreline wave-power system. Ocean Engineering. 2000;**27**(8):801-821

[33] Falcao OAF. The shoreline OWC wave power plant at the Azores. In: Proceedings of the Fourth European Wave Energy Conference. Aalborg, Denmark: Aalborg University Publisher; 2000. pp. 42-48

[34] Kofoed JP, Frigaard P, Madsen EF, Soransen HC. Prototype testing of the wave energy converter wave dragon. Renewable Energy. 2006;**31**(2):181-189

[35] Li X, Martin D, Jiang B, Chen S, Thiagarajan K, Parker RG, et al. Analysis and wave tank verification

of the performance of point absorber WECs with different configurations. IET Renewable Power Generation. 2021;**15**(14):3309-3318

[36] Lobato P, Cruz A, Silva J, Pires AJ. The switched reluctance generator for wind power generation. In: Proceedings of the IEEE PES Conference Asociación Española para el Desarrollo de la Ingeniería Eléctrica. Piscataway, New Jersey, USA: IEEE; 2009

[37] Presivic M. Wave power technologies (paper 05GM0542). In: Proceedings of the IEEE Power Engineering Society 2005 Meeting Panel Session. Palo Alto, CA: Electric Power Research Institute; 2005

[38] Polinder H, Scuotto M. Wave energy converters and their impact on power systems. In: Proceedings of the International Conference on Future Power Systems. Piscataway, New Jersey, USA: IEEE; 2005. pp. 1-9

[39] Valerio D, Beirao P, da Costa JS. Optimization of wave energy extraction with the Archimedes wave swing. Ocean Engineering. 2007;**34**(17-18):2330-2344

[40] Reference Model Project (RMP): Sandia Energy. Available from: https://energy.sandia.gov/programs/renewable-energy/water-power/projects/reference-model-project-rmp/

[41] Aquamarine Power: EMEC: European Marine Energy Centre. Available from: https://www.emec.org.uk/about-us/wave-clients/aquamarine-power/

[42] Li Q, Mi J, Li X, Chen S, Jiang B, Zuo L. A self-floating oscillating surge wave energy converter. Energy. 2021;**230**:120668

[43] Evans DV. The oscillating water column wave-energy device. IMA Journal of Applied Mathematics. 1978;**22**(4):423-433

[44] Engström J, Kurupath V, Isberg J, Leijon M. A resonant two body system for a point absorbing wave energy converter with direct-driven linear generator. Journal of Applied Physics. 2011;**110**(12):124904

[45] Kolios A, Di Maio LF, Wang L, Cui L, Sheng Q. Reliability assessment of point-absorber wave energy converters. Ocean Engineering. 2018;**163**:40-50

[46] Li X, Chen CA, Chen S, Xiong Q, Huang J, Lambert S, et al. Dynamic characterization and performance evaluation of a 10-kW power take-off with mechanical motion rectifier for wave energy conversion. Ocean Engineering. 2022;**15**(250):110983-110983

[47] Li X, Chen C, Li Q, Xu L, Liang C, Ngo K, et al. A compact mechanical power take-off for wave energy converters: Design, analysis, and test verification. Applied Energy. 2020;**278**:115459

[48] Yang L, Huang J, Congpuong N, Chen S, Mi J, Bacelli G, et al. Control co-design and characterization of a power takeoff for wave energy conversion based on active mechanical motion rectification. IFAC-PapersOnLine. 2021;**54**(20):198-203

[49] Chen S, Jiang B, Li X, Huang J, Wu X, Xiong Q, et al. Design, dynamic modeling and wave basin verification of a hybrid wave–current energy converter. Applied Energy. 2022;**321**:119320

[50] Waveenergy.it. The First Worldwide Application of Full-Scale OWC at the Port of Civitavecchia. Available from: https://www.wavenergy.it/generico/

the-first-worldwide-application-at-full-scale-of-the-rewec3-device-in-the-port-of-civitavecchia/

[51] Ocean Power Technologies. PB3 PowerBuoy®. Available from: https://oceanpowertechnologies.com/pb3-powerbuoy

[52] Cruz J. Ocean Wave Energy: Current Status and Future Perspectives. Berlin: Springer; 2018

[53] Makai Ocean Engineering. Ocean Thermal Energy Conversion. Available from: https://www.makai.com/ocean-thermal-energy-conversion/

[54] Nihous GC. Mapping available ocean thermal energy conversion resources around the main Hawaiian islands with state-of-the-art tools. Journal of Renewable and Sustainable Energy. 2020;**12**(1):014501

[55] Straub AP, Deshmukh A, Elimelech M. Pressure-retarded osmosis for power generation from salinity gradients: Is it viable? Energy & Environmental Science. 2016;**9**(1):31-48

[56] Reuters. Norway Opens World's First Osmotic Power Plant. Available from: https://www.reuters.com/article/us-norway-osmotic-idUSTRE5AN20Q20091124/

[57] REDStack. Reverse Electrodialysis Technology. Available from: https://www.redstack.nl/

[58] Reuters. Sunseap to Build World's Largest Floating Solar Farm and Energy Storage System. Available from: https://www.reuters.com/business/energy/sunseap-build-2-bln-floating-solar-farm-indonesia-worlds-largest-2021-07-22/#:~:text=SINGAPORE%2C%20July%2022%20(Reuters),its%20renewable%20power%20generation%20capacity

[59] PV Tech. Oceans of Energy to Build First Commercial Offshore Solar-Wind Project in North Sea. Available from: https://www.pv-tech.org/oceans-of-energy-to-build-worlds-first-commercial-offshore-solar-project-in-north-sea/

[60] Algenol Biotech LLC. Algenol Biotech LLC - Algae Technology. Available from: https://www.algenol.com/

[61] MacroFuels. MacroFuels Project: Developing the Next Generation Biofuels from Seaweed. Available from: http://www.macrofuels.eu/

[62] Mérigaud A, Ringwood JV. Condition-based maintenance methods for marine renewable energy. Renewable and Sustainable Energy Reviews. 2016;**66**:53-78

[63] Gracia S, Olivito J, Resano J, Martin-del-Brio B, de Alfonso M, Álvarez E. Improving accuracy on wave height estimation through machine learning techniques. Ocean Engineering. 2021;**236**:108699

[64] Forootan MM, Larki I, Zahedi R, Ahmadi A. Machine learning and deep learning in energy systems: A review. Sustainability. 2022;**14**(8):4832

[65] Williamson BJ, Blondel P, Armstrong E, Bell PS, Hall C, Waggitt JJ, et al. A self-contained subsea platform for acoustic monitoring of the environment around marine renewable energy devices–field deployments at wave and tidal energy sites in Orkney, Scotland. IEEE Journal of Oceanic Engineering. 2015;**41**(1):67-81

[66] Copping AE, Hemery LG, Overhus DM, Garavelli L,

Freeman MC, Whiting JM, et al. Potential environmental effects of marine renewable energy development—The state of the science. Journal of Marine Science and Engineering. 2020;**8**(11):879

[67] International Renewable Energy Agency. Renewable Power Generation Costs in 2022. Abu Dhabi: IRENA; 2023

[68] Simbolon R, Sihotang W, Sihotang J. Tapping ocean potential: Strategies for integrating tidal and wave energy into national power grids. GEMOY: Green Energy Management and Optimization Yields. 2024;**1**(1):49-65

[69] Mwasilu F, Jung JW. Potential for power generation from ocean wave renewable energy source: A comprehensive review on state-of-the-art technology and future prospects. IET Renewable Power Generation. 2019;**13**(3):363-375

[70] Buck BH, Langan R. Aquaculture Perspective of Multi-Use Sites in the Open Ocean: The Untapped Potential for Marine Resources in the Anthropocene. Cham, Switzerland: Springer Nature; 2017

[71] Khawaji AD, Kutubkhanah IK, Wie JM. Advances in seawater desalination technologies. Desalination. 2008;**221**(1-3):47-69

[72] Vicinanza D, Margheritini L, Kofoed JP, Buccino M. The SSG wave energy converter: Performance, status and recent developments. Energies. 2012;**5**(2):193-226

[73] Ayers JM, Richter K. The potential of small-scale turbines and microbial fuel cells to support persistent oceanographic sensors. In: OCEANS 2016 MTS/IEEE Monterey Conference is Piscataway. New Jersey, USA: IEEE; 2016. pp. 1-6

[74] Yates KL, Schoeman DS, Klein CJ. Ocean zoning for conservation, fisheries and marine renewable energy: Assessing trade-offs and co-location opportunities. Journal of Environmental Management. 2015;**152**:201-209

[75] Sherman P, Van Sebille E. Modeling marine surface microplastic transport to assess optimal removal locations. Environmental Research Letters. 2016;**11**(1):014006

[76] United Nations Sustainable Development Goals. Available from: https://www.sdgs.un.org/goals

Chapter 6

# Grid-Connected Photovoltaic Systems with Filtering and Reactive Power Injection

*Abdelmajid Abouloifa, Meriem Aourir, Chaouqi Aouadi and Ibtissam Lachkar*

## Abstract

This chapter investigates the control of a shunt active power filter (SAPF) integrated with a solar photovoltaic (PV) panel to meet stringent load requirements, encompassing the delivery of active, reactive, and harmonic power with a primary focus on power factor correction at the grid side. A key contribution of this study is the augmentation of the system with the capability to inject supplementary reactive power into the grid, thereby enhancing grid stability. The adopted control strategy utilizes the flatness approach, which was selected for its efficacy in managing complex systems. This methodology ensures efficient coordination of power flow, optimizing the injection of reactive power while ensuring grid synchronization and stability. Detailed simulation results substantiate the effectiveness of the proposed system in achieving enhanced power quality and grid performance. This chapter contributes to advancing the understanding and implementation of hybrid power systems by demonstrating the integration of renewable energy sources with advanced control strategies. By enhancing grid stability and efficiency, this research underscores the potential of SAPF-PV systems in modern power grids.

**Keywords:** shunt active power filter (SAPF), flatness control approach, reactive power injection, grid stability and synchronization, MPPT

## 1. Introduction

Given the escalating global electricity needs, solar photovoltaic (PV) energy stands out as a crucial renewable energy source. Its importance lies in its vast availability, the declining costs of PV technology, and its environmentally friendly nature. Solar PV systems offer a sustainable solution to meet energy demands while significantly reducing greenhouse gas emissions and dependence on fossil fuels [1, 2].

Beyond power generation, the integration of active power filters (APFs) with PV systems plays crucial role in enhancing power quality. APFs mitigate power quality issues such as harmonics and reactive power, which are common in electrical grids [3]. This integration ensures that the electricity produced by PV systems is not only sustainable but also stable and of high quality. Consequently, the synergy between solar PV energy and active power filtering contributes to a more reliable, efficient, and environmentally friendly energy infrastructure, addressing both the demand for clean energy and the necessity for high power quality in modern electrical grids [4].

In parallel, the efficiency of buildings' energy use is crucial for reducing energy dependence and environmental impact. Controlling energy flows within buildings is vital for occupant comfort and timely use of consumer goods. Recent studies have developed ensemble machine learning techniques for heat loading (HL) and cool loading (CL) estimation in intelligent urban ecosystems. This advancement underscores the importance of adopting intelligent systems to optimize energy use in buildings, complementing the role of solar PV systems and APFs in the broader energy landscape [5].

### 1.1 Introduction to power factor

Power factor is a critical parameter in electrical systems, representing the ratio of real power, which performs useful work, to apparent power, which is the combination of real power and reactive power [6]. It is a measure of how effectively electrical power is being used. A power factor close to unity indicates efficient utilization of electrical power, whereas a lower power factor signifies inefficiency. Reactive power, which does not perform any useful work but is necessary to maintain voltage levels for the transfer of active power, contributes to this inefficiency. Poor power factor can arise from various sources, such as inductive loads like motors and transformers, which consume reactive power [7].

- *Power factor for linear loads*:

In the context of linear loads, the apparent power, denoted as kVA ($S = V \times I$), is the vector combination of reactive power Q and real power P as shown in **Figure 1**. The power factor is calculated as:

$$PF = \frac{|P|}{S} = \frac{1}{\sqrt{1+\frac{Q^2}{P^2}}} = \cos(\Phi)$$

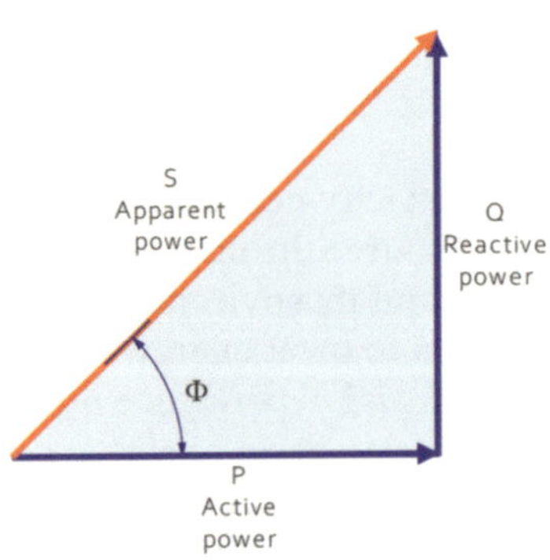


**Figure 1.**
*Power triangle in linear loads.*

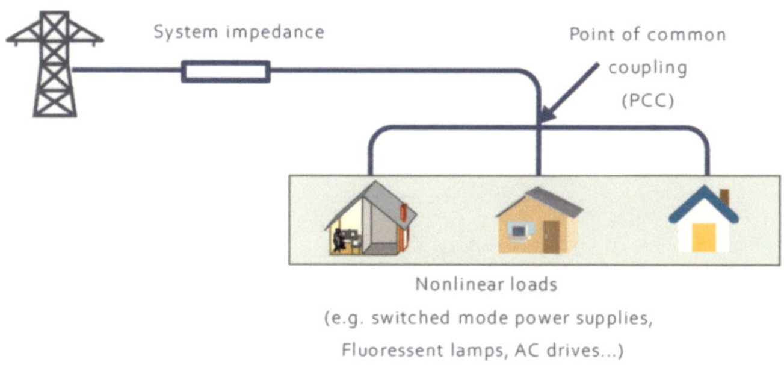


**Figure 2.**
*Power grid configuration with nonlinear loads at the point of common coupling (PCC): a schematic overview*

where Φ is the phase angle between S and P. This phase angle is equivalent to the displacement angle between the voltage and current in linear loads. When the displacement angle increases for a given current, reactive power Q rises, real power P falls, and the power factor diminishes. Inductive loads, like induction motors, make the current lag behind the voltage. Conversely, capacitors make the current lead the voltage, and purely resistive loads align the current with the voltage phase. For systems that solely have linear loads (a rare scenario), adding capacitor banks can help improve a lagging power factor caused by induction motors or other lagging loads (**Figure 2**), [8].

- *Power factor for nonlinear loads*:

Nonlinear loads, such as electronic devices with switching power supplies, create harmonic currents that deviate from the ideal sinusoidal waveform. Indeed, nonlinear loads contribute reactive power without performing useful work. This modifies the power vector into a three-dimensional model where distortion reactive power combines with reactive and real power to form the system's apparent power [9]. The power factor, the ratio of kW to kVA, now includes a harmonic component (see **Figure 3**), making the true power factor a mix of displacement and distortion factors. Though displacement power factors are near unity for nonlinear loads, the true power factor is usually low due to distortion, such as a personal computer with a power factor around 0.65 to 0.7. Eliminating harmonic currents is the best way to improve power factor [10].

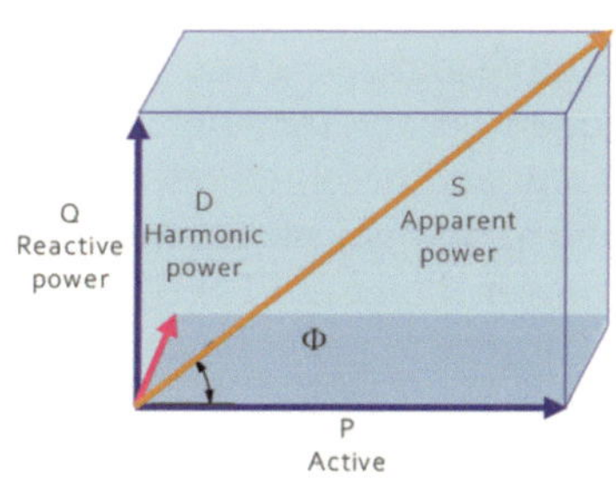


**Figure 3.**
*Power triangle in nonlinear loads.*

$$
\begin{aligned}
PF &= \frac{1}{\sqrt{1+\dfrac{Q^2+D^2}{P^2}}} \\
&= \frac{1}{\sqrt{1+\dfrac{\sum_{n=2}^{\infty} I_n^2}{I_1^2}}} \times \cos(\phi)
\end{aligned}
$$

### 1.2 Challenges of a poor power factor

The integration of a wide range of power electronic converters introduces nonlinear dynamic loads that pose instability risks for power grids. These converters produce higher-order harmonics and reactive currents that distort the sinusoidal form of the alternating current. These distortions propagate to other loads connected to the point of common coupling, with harmful effects on power grid stability and distribution quality [11]. As a matter of fact, a poor power factor in electrical systems presents several significant challenges. Primarily, it leads to increased losses in the distribution network due to higher currents, which in turn cause greater heat dissipation in conductors and transformers. This inefficiency necessitates the use of larger conductors and transformers to handle the additional current, leading to higher infrastructure costs. Additionally, a low power factor reduces the overall capacity of the electrical system, as a larger portion of the current is non-productive reactive power. This can result in higher electricity bills for consumers due to penalties imposed by utility companies for maintaining a power factor below a specified threshold. Moreover, the presence of a poor power factor can cause voltage drops and stability issues, which contribute to significant deterioration of power quality and affect the effective performance of consumers' devices connected to the same point of common coupling (PCC). Addressing power factor problems is therefore crucial for optimizing energy efficiency, reducing operational costs, and ensuring the reliable performance of electrical systems.

### 1.3 Enhancing power factor: effective solutions and strategies

Passive filters are commonly employed to improve power factor by mitigating harmonic distortion and reactive power in electrical systems. These filters, typically consisting of inductors, capacitors, and resistors, are designed to target specific harmonic frequencies, thereby reducing their impact on the overall power quality. While passive filters are relatively simple and cost-effective, they have notable drawbacks. They are limited to fixed frequencies and may not adapt well to varying load conditions, leading to resonance issues and potential overloading [12]. Additionally, their effectiveness diminishes as the harmonic spectrum of the load changes.

In contrast, shunt active power filters (APFs) offer a more dynamic and versatile solution. APFs use power electronics to generate compensating currents that counteract harmonic and reactive components in real time, thus improving power factor and overall power quality across a wide range of operating conditions. Unlike passive filters, APFs can adapt to varying loads and harmonic profiles, providing consistent performance without the risk of resonance. Although they are more complex and expensive than passive filters, the advantages of shunt active power filters, including their adaptability, precision, and ability to address multiple harmonic orders simultaneously, make them a superior choice for maintaining optimal power quality in modern electrical systems [13].

### 1.4 Integration of PV systems with shunt active power filtering for enhanced power factor correction

A hybrid system that integrates photovoltaic (PV) energy with shunt active power filtering capabilities represents a sophisticated solution for modern power grids. Such a system not only injects active power generated from PV panels into the electrical grid but also performs power factor correction to enhance overall efficiency. The active power generated by the PV system helps meet the energy demand while reducing dependence on conventional power sources, promoting sustainable energy use [14]. Simultaneously, the shunt active power filter compensates for reactive power and harmonics caused by nonlinear loads, ensuring that the power factor remains close to unity [15]. This dual functionality is achieved through advanced power electronics and control strategies that dynamically adjust to the varying conditions of both the PV generation and the load demands. By addressing both energy production and power quality, this integrated approach improves grid stability, reduces energy losses, and enhances the reliability of the power supply.

## 2. Single-phase grid-connected PV systems with active filtering and reactive power injection capability

In modern power systems, the integration of single-phase grid-connected PV systems plays a crucial role in enhancing energy efficiency and sustainability. These systems are designed not only to inject active power generated from solar energy into the grid but also to perform ancillary services such as active filtering and reactive power injection. Active filtering is essential for mitigating harmonics and improving power quality by eliminating unwanted frequencies from the power supply. Reactive power injection, on the other hand, supports voltage stability and improves the overall power factor of the grid. By incorporating these capabilities, single-phase PV systems contribute significantly to the stability and reliability of the electrical grid. Advanced control strategies and inverter technologies are employed to ensure seamless integration and optimal performance of these systems, addressing both the dynamic demands of modern power networks and the intermittent nature of solar energy [16, 17]. As a result, these systems represent a multifaceted approach to achieving a cleaner, more resilient, and efficient power infrastructure.

### 2.1 System presentation

The studied system that can manage power quality and support grid stability is illustrated in **Figure 4**. It consists of three principal parts: a single-phase electrical grid, a single-phase nonlinear loads, and a PV panel connected to a two-level IGBT-based full-bridge SAPF with filtering impedance $(r_f, L_f)$ and a capacitor for energy storage $(C_{pv})$ placed on the DC side. The sinusoidal grid voltage is given as:

$$v_g = \sqrt{2} V_g \sin(\omega_g t) \tag{1}$$

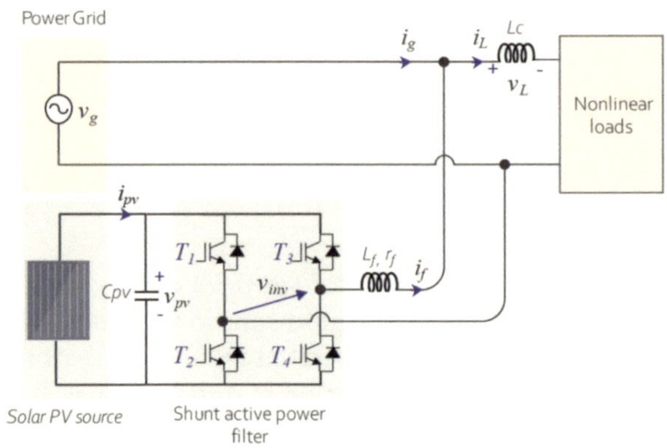


**Figure 4.**
*Circuit of the PV-SAPF association.*

where $V_g$ and $\omega_g$ stand for the RMS value and the angular frequency of the grid voltage, respectively. The current through nonlinear loads, $i_L(t)$, is characterized by its complex waveform, which deviates from a pure sinusoidal form due to the presence of harmonic distortions. This non-sinusoidal nature of the current can be described by its Fourier expansion, expressed as:

$$i_L(t) = \sum_{h=1}^{\infty} I_{L,h} \sin\left(h\omega_g t + \varphi_h\right) \tag{2}$$

where $i_{L,h}$ represents the amplitude and $\varphi_h$ the phase angle of the $h - th$ harmonic component. These harmonics are generated by the nonlinear elements, such as rectifiers and switching devices.

## 2.2 System mathematical modeling

This subsection provides the mathematical formulations and equations that describe the dynamics of the system. Indeed, the switching function $u$ of the inverter taking values in the finite set $\{-1, 1\}$ as follows:

$$u = \begin{cases} 1 \text{if } (T_1, T_4) \text{ are ON and } (T_2, T_3) \text{ are OFF} \\ -1 \text{if } (T_1, T_4) \text{ are ON and } (T_2, T_3) \text{ are OFF} \end{cases}$$

Using Kirchhoff's laws, the instantaneous behavior of the single-phase SAPF-PV system can be represented by the following differential equations:

$$C_{pv} \frac{dv_{pv}}{dt} = i_{pv} - u\ i_f \tag{3}$$

$$L_f \frac{di_f}{dt} = -r_f i_f + u\ v_{pv} - v_g \tag{4}$$

The instantaneous model Eq. (3) is used for accurate numerical simulations, but its direct use in the development of nonlinear controllers is restricted by the inclusion of binary signals. Consequently, the following averaged model is adopted:

$$C_{pv}\frac{d\bar{v}_{pv}}{dt} = i_{pv} - D\ \bar{i}_f \tag{5}$$

$$L_f\frac{d\bar{i}_f}{dt} = -r\bar{i}_f + D\ \bar{v}_{pv} - \bar{v}_g \tag{6}$$

where $\bar{v}_{pv}$, $\bar{i}_f$, $D$, and $\bar{v}_g$ are the averaged values over one switching period of the instantaneous signals $v_{pv}$, $i_f$, $u$, and $v_g$.

## 3. Controller design

In this section, we develop a control strategy grounded in flatness theory, leveraging the inherent flatness property of the studied system. By utilizing the average model, this controller is meticulously designed to exploit the system's flatness property, enabling precise trajectory planning and control. The structure of the designed controller features two cascaded loops, as shown in **Figure 5**. The proposed control approach is particularly advantageous for achieving the primary objectives of the controller in PV systems, which are crucial for optimizing performance and ensuring stability in power transmission. First, Maximum Power Point Tracking (MPPT) is essential to maximize the extraction of available power from the PV array. This involves dynamically adjusting the operating point of the PV panel to ensure it operates at its optimal power output under varying environmental conditions. Secondly, the controller must manage the reactive ($Q$) and harmonic ($D$) power compensation for nonlinear loads. Lastly, the controller facilitates the injection of supplementary reactive power ($Qs$) into the grid. This function is vital for supporting

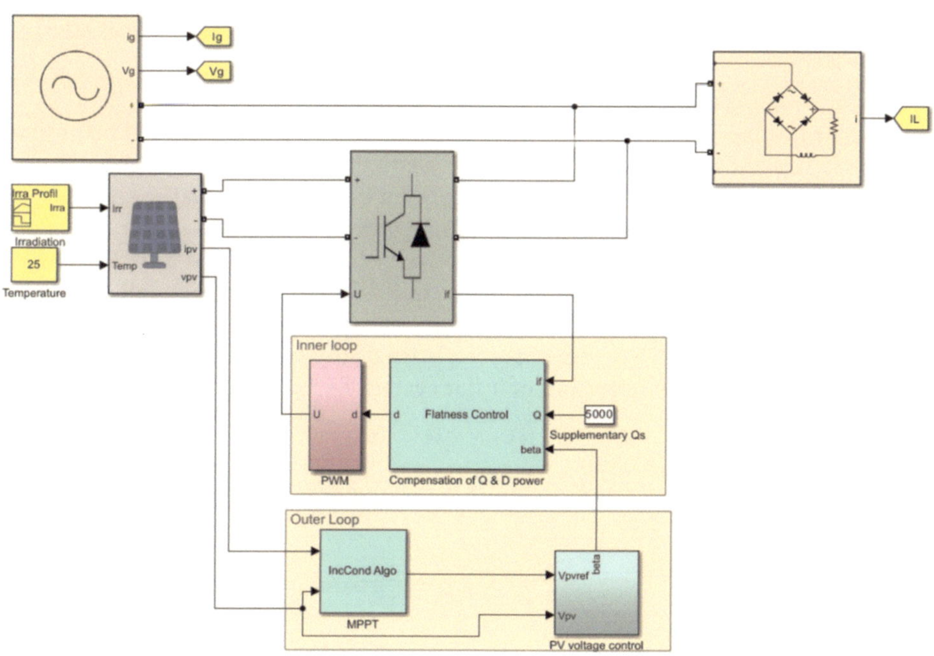


**Figure 5.**
*Overall simulation diagram of the PV-SAPF system.*

the voltage levels and improving the power factor of the grid, thereby enhancing the overall reliability and performance of the electrical grid. By achieving these objectives, the controller ensures optimal power extraction, improved power quality, and enhanced grid stability.

### 3.1 Inner loop design: reactive and harmonic power compensation with supplementary Qs injection

To ensure the compensation of reactive and harmonic power drawn by nonlinear loads and inject a supplementary reactive power into the grid, the grid current $\bar{i}_g$ should track a reference signal expressed as follows:

$$\bar{i}_g^{ref} = \beta v_g(t) - \gamma v_g(t - T/4) \tag{7}$$

where

$$\beta \propto P_g \quad \text{and} \quad \gamma \propto Q_s$$

$\beta$ serves as a dynamic conductance parameter that determines the amplitude of the grid current, crucial for regulating the power flow. Meanwhile, $\gamma$, derived as $\gamma = Q_s/V_g^2$, dictates the level of supplementary reactive power to be injected into the power grid. This requirement is met if the filter current injected by SAPF matches the following reference signal:

$$\bar{i}_f^{ref} = i_L - \bar{i}_g^{ref} \tag{8}$$

Consider the following tracking error, associated with current compensation:

$$e_i = \bar{i}_f - \bar{i}_f^{ref} \tag{9}$$

Using the principles of differential flatness, the inner control loop will be designed. This approach is chosen for its effectiveness in situations requiring the generation of explicit trajectories. Flatness theory aims to produce intrinsically flat outputs corresponding to the dimensions of the input signals of the system under consideration [18]. In the realm of differential algebra, a system is classified as flat if there exists a specific set of variables termed flat outputs. These outputs enable the determination of all state variables and input components without the need for integration. Prior to demonstrating the flatness characteristics of the system under consideration, it is essential to revisit the definition of a flat system defined by state $x$ and input $u$:

$$\dot{x} = f(x, u) \quad x \in \mathbb{R}^n, \quad u \in \mathbb{R}^m \tag{10}$$

where $\dot{x}$ is a flat system with state $x$ and input $u$, if there is a flat output $y \in \mathbb{R}^m$ such that:

$$y = \phi(x, u, \dot{u}, \dots, u^r) \tag{11}$$

Projecting this into the system under study implies the obligation of defining one flat output, which is chosen as follows:

$$y = \bar{i}_f \tag{12}$$

This choice results in achieving a system that is differentially flat. By applying equations Eq. (6) and Eq. (12), one can derive:

$$D = \frac{L_f \dot{y} + r_f \ y + \bar{v}_g}{\bar{v}_{pv}} \tag{13}$$

Before formulating the control law, it is necessary to design the desired trajectory for the flat output. To achieve this, a first-order low-pass filter is applied to the reference signal. Thus, the reference trajectory can be expressed as follows:

$$y^p = \frac{1}{1 + \tau s} y^{ref} \tag{14}$$

where $\tau$ is time constant of the filter. Now, the controller needs to be designed to ensure that the flat output signal $y$ tracks its planned trajectory $y_p$. Therefore, the following state feedback controller is proposed:

$$\left(\dot{y}_p - \dot{y}\right) + K\left(y_p - y\right) = 0 \tag{15}$$

The design parameter $K$ is selected to be a positive real number. The control input $D$ can then be computed using equation Eq. (15) to produce the binary control signal $u$ for the SAPF-PV system, achieved via a PWM generator.

### 3.2 Outer loop design: MPPT objective

The aim of this part is to design a controller that can effectively manage the power distribution between photovoltaic panels, nonlinear loads, and the power grid. To achieve this, a type II controller, which combines a Proportional-Integral (PI) controller with a low-pass filter, is developed to generate an additional control signal, $\beta$. This advanced controller ensures that the square of the photovoltaic (PV) voltage, denoted as $z = \bar{v}_{pv}^2$, closely tracks the setpoint $z^* = \left(v_{pv}^{ref}\right)^2$ as determined by the Maximum Power Point Tracking (MPPT) block.

#### *3.2.1 Implementation of an IncCond algorithm for enhanced MPPT performance in PV systems*

The IncCond algorithm is used to ensure the Maximum Power Point Tracking (MPPT) in photovoltaic systems. This algorithm operates by continuously adjusting the voltage to maximize the power output. The fundamental principle involves comparing the incremental conductance ($\frac{di_{pv}}{dv_{pv}}$) to the instantaneous conductance ($\frac{i_{pv}}{v_{pv}}$). When these two values are equal, the maximum power point (MPP) is achieved, and the algorithm maintains this voltage to ensure optimal performance. If there are changes in current, the algorithm dynamically adjusts the voltage, either increasing or decreasing it, to track the new MPP efficiently [19].

### 3.2.2 PV voltage regulation using PI type II controller

Let control law Eq. (13) be applied to the system of **Figure 4**, here described by the average model Eq. (4), and it can be demonstrated that the squared PV voltage $z = \overline{v}_{pv}^2$ is related to the tuning signal $\beta$, according to the following time-varying linear equation:

$$\frac{dz}{dt} = \frac{2P_{pv}}{C_{pv}} - f\left(t, e_i, \beta, \dot{\beta}\right) \tag{16}$$

where

$$f\left(t, e_i, \beta, \dot{\beta}\right) = \frac{1}{C_{pv}}\left[\left(\sum_{h=1}^{\infty} i_{L,h}\sin\left(h\omega_g t + \varphi_h\right) - \beta V_g \sin\left(\omega_g t\right)\right)\right.$$
$$\left[\left(1 - r_f\beta - L_f\dot{\beta}\right)V_g \sin\left(\omega_g t\right) - L_f\beta\omega_g\cos\left(\omega_g t\right) + \right.$$
$$\left.\left. r_f\sum_{h=1}^{\infty} i_{L,h}\sin\left(h\omega_g t + \varphi_h\right) + L_f\sum_{h=1}^{\infty} i_{L,h}h\omega_g\cos\left(h\omega_g t + \varphi_h\right)\right]\right]$$

The signal $\beta$ acts as a control input for the system described by Eq. (16). Our objective is to establish a tuning law for $\beta$ such that the squared voltage $z = (\overline{v}pv)^2$ accurately follows a given reference signal $z^{ref} = \left(vpv^{ref}\right)^2$, which is provided by the IncCond algorithm. Considering the necessity for the first derivative of $\beta$ to be accessible, the following type II controller, which integrates a Proportional-Integral (PI) controller with a low-pass filter, is employed to achieve this goal.

$$\beta(s) = \frac{c_1}{c_1 + s}\left(k_p\hat{e}_1(s) + k_i\hat{e}_2(s)\right) \tag{17}$$

where $e_1$ is the error between the variable $y$ and its reference $y^{ref}$ and $e_2$ is its time integral.

$$e_1 = z^{ref} - z \quad \text{and} \quad e_2 = \int_0^t e_1(\tau)d\tau \tag{18}$$

The over hat in Eq. (17) stands for taking the Laplace transform and $\left(c_1, k_p, k_i\right)$ are suitable positive real design parameters. Using Eq. (17), the first time-derivative of $\beta$ is given by:

$$\dot{\beta}(t) = c_1\left(k_p e_1 + k_i e_2 - \beta(t)\right) \tag{19}$$

## 4. Results and discussion

For the simulation process, the single-phase full-bridge SAPF-PV system was numerically evaluated in MATLAB/Simulink using the proposed method based on a flatness controller. The detailed parameters used by the power system are summarized in **Table 1**. To test the robustness of the closed-loop control system performance,

| Parameter | Symbol | Value |
|---|---|---|
| Electrical Grid | $V_g, \Omega_g$ | $230V, 50Hz$ |
| PV-SAPF | $N_s$ | 30 |
| | $N_p$ | 1 |
| | $C_{pv}$ | $10mF$ |
| | $r_f, L_f$ | $8m\Omega, 3mH$ |
| | $f_s$ | $10kHz$ |
| Nonlinear load | $L_c, R, L$ | $1mH, 5\Omega, 500mH$ |
| PV voltage controller | $k_p, k_i, c_1$ | $0.01, 0.002, 8e^{-2}$ |
| Current controller | $K$ | $10e^5$ |

**Table 1.**
*Global system parameters including plant and controller.*

some important operating modes (OP) are considered and described as follows: The performance of the power system, including its control unit, will now be evaluated by simulation using MATLAB/Simulink. To check the performance of the system in a closed loop, two simulation scenarios were considered; the first one, two operating modes have been described as follows: OP1: Under standard climatic conditions ($G = 1000W/m^2$ and $T = 25degC$), the available PV power is insufficient ($P_{pv}$ inf $P_L$), requiring the electrical grid to supply the active power demanded by the nonlinear load ($P_L = P_{pv} + P_g$). OP1:n this mode, the PV panel delivers its maximum power ($P_{pv} = 8.75kW$), with the excess power ($P_{pv} - P_L$) being entirely fed into the electrical grid.

### 4.1 Simulation results under varying climatic conditions

The first operating mode condition is shown in **Figures 6–12**. It is clear from **Figures 6** and 7 that the PV voltage follows its maximum value and consequently, the generated power from the PV panel follows its maximum value $\left(P_{pv} = 8.75kW\right)$. Furthermore, **Figure** 7 shows that the power feeding the load is coming from the PV and the grid.

**Figure 8** illustrates the waveform of the output filter current that ensures the compensation of the harmonic currents drawn by the nonlinear loads. **Figure 9** shows

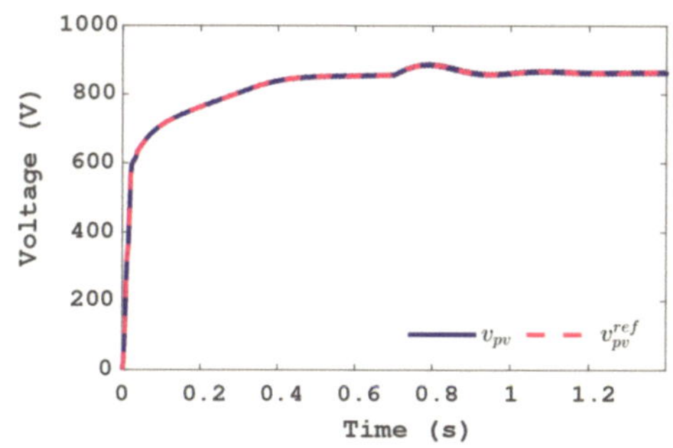


**Figure 6.**
*PV voltage with its optimal reference provided by InCond algorithm under varying irradiance.*

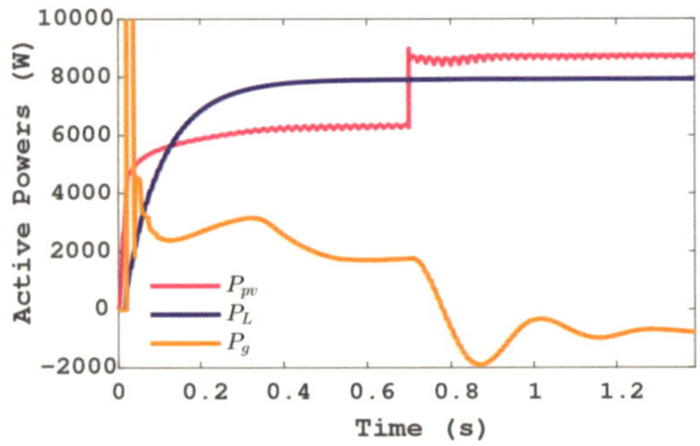


**Figure 7.**
*Output PV, load, and grid powers under varying irradiance.*

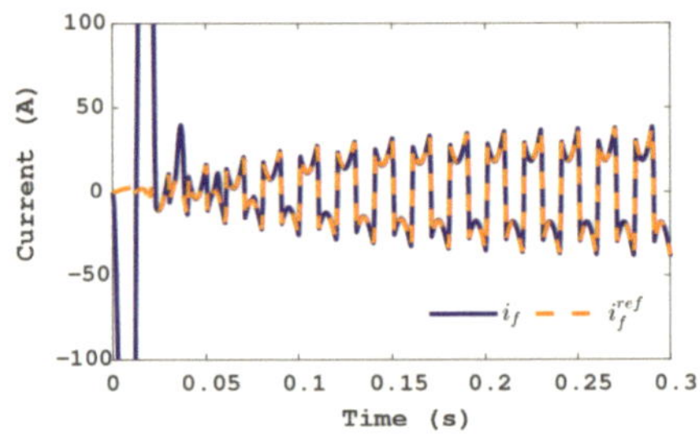


**Figure 8.**
*Output filter current tracking its reference.*

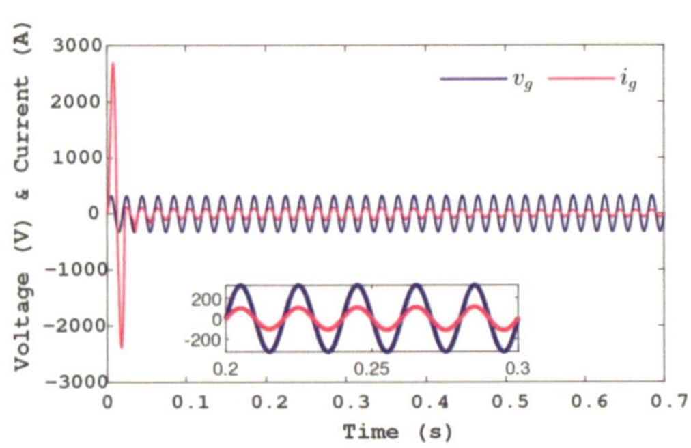


**Figure 9.**
*Grid voltage and current waveforms.*

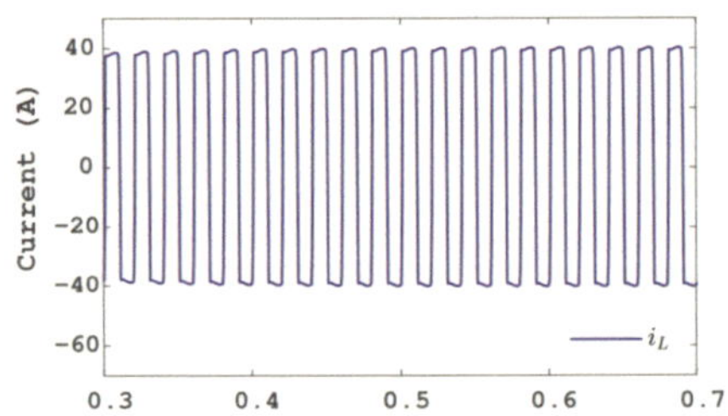


**Figure 10.**
*Load current waveform.*

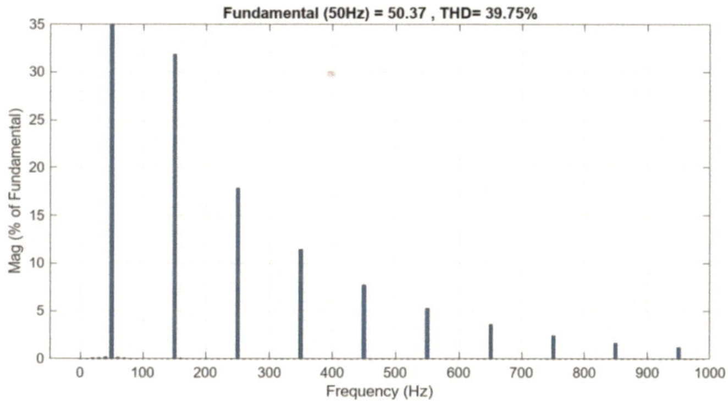


**Figure 11.**
*FFT analysis of the load current.*

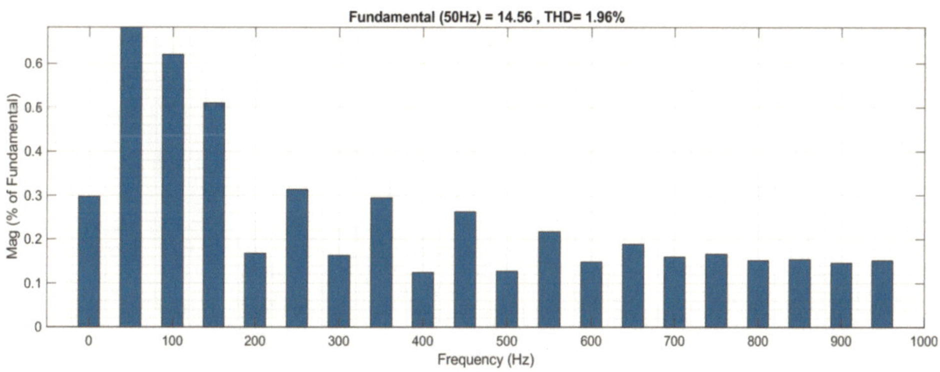


**Figure 12.**
*FFT analysis result of the grid current.*

good compensation of harmonic and reactive currents, so that the grid current is sinusoidal and in phase with the grid voltage. **Figure 10** highlights that the inductive load current is strongly perturbed and this is confirmed by the result of its FFT analysis which indicates a THD of 39.75% (see **Figure 11**). Finally, **Figure 12** displays the THD value in the grid current and shows a value of 1.86%, demonstrating the effectiveness of the proposed system and control strategy.

### 4.2 Simulation results considering reactive power injection into the grid

In this scenario, the injection of supplementary reactive power into the grid is considered in order to support the grid stability. The results that prove the effectiveness of the proposed control strategy are shown in **Figures 13–16**. **Figure 13** illustrates the PV voltage which tracks with good accuracy its setpoint provided by the MPPT algorithm. **Figure 14** presents the output PV, load, and grid active powers. **Figure 15** highlights the effect of reactive power injection on filter current. Finally, from **Figure 16**, it can be seen that the amplitude of the grid current is changed and that there is a phase shifting between grid voltage and current, which confirms the ability of the system to inject reactive power into the electrical grid.

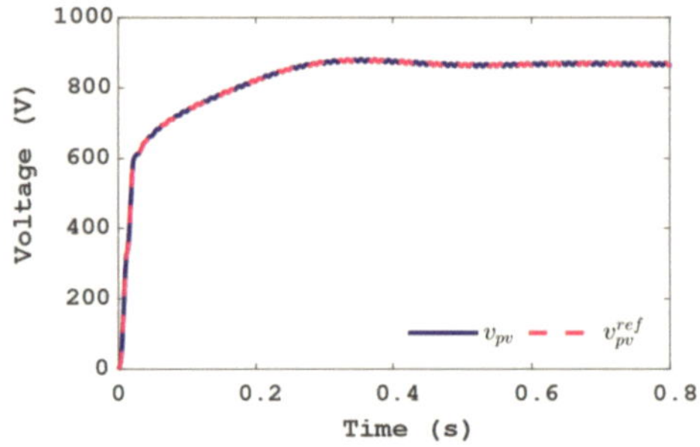


**Figure 13.**
*PV voltage considering reactive power injection into the grid.*

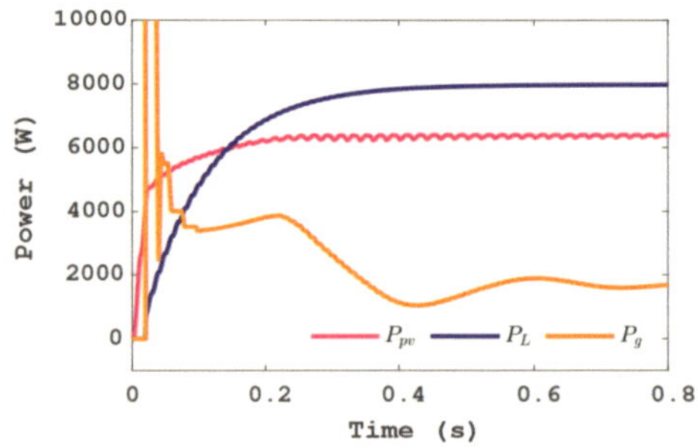


**Figure 14.**
*Output PV, load, and grid power considering reactive power injection into the grid.*

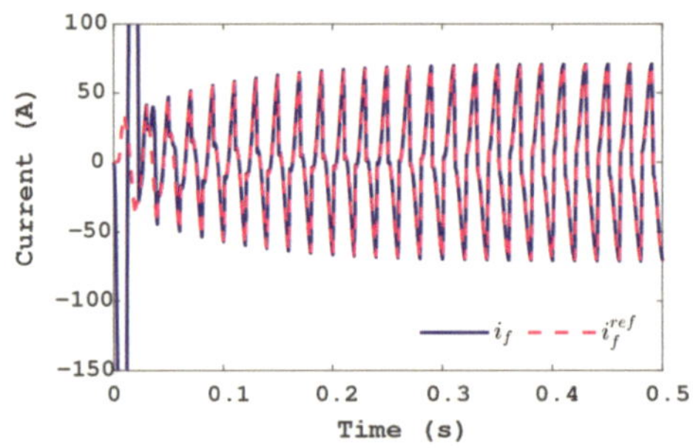


**Figure 15.**
*Output filter current in the presence of reactive power injection into the grid.*

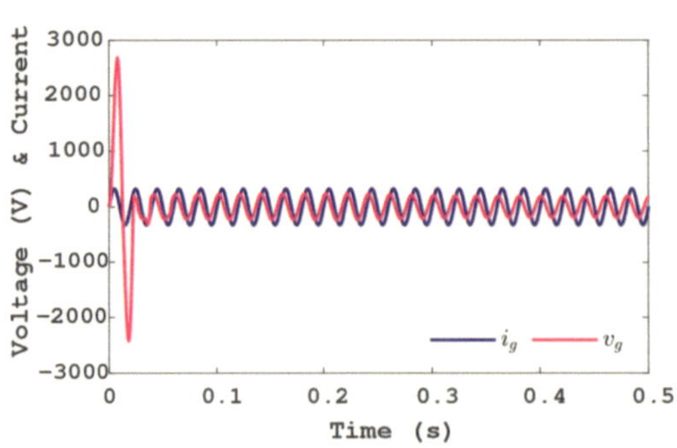


**Figure 16.**
*Voltage and grid current waveforms under reactive power injection.*

## 5. Conclusions

In this chapter, we explored the innovative concept of using a shunt active power filter (SAPF) fed by a solar PV panel to effectively manage load requirements and enhance grid stability. This approach addresses the demands for active, reactive, and harmonic power while also offering the unique advantage of injecting supplementary reactive power into the grid, thus aiding in stabilizing the grid and meeting the requirements of other loads. Our control strategy based on flatness control approach, and the obtained results demonstrate the effectiveness of this method in ensuring the system's global requirements. The flatness approach proved to be a robust control strategy, capable of maintaining the desired performance of the SAPF and optimizing the interaction with the solar PV system. By integrating SAPF with solar PV technology, we present a sustainable and efficient solution for modern power systems. Harnessing renewable energy not only reduces reliance on conventional power sources but also contributes to a greener, more resilient grid infrastructure. The dual functionality of the system meeting immediate load demands and supporting grid stability underscores its potential as a key component in future smart grid developments.

## Conflict of interest

The authors declare no conflict of interest.


## Author details

Abdelmajid Abouloifa[1†], Meriem Aourir[2*], Chaouqi Aouadi[1†] and Ibtissam Lachkar[1]

1 ESE Lab, ENSEM of Casablanca, Hassan II University, Casablanca, Morocco

2 LIME Lab, Higher Institute of Engineering and Business ISGA, Casablanca, Morocco

*Address all correspondence to: meriem.aourir@isga.ma

† These authors contributed equally.

## References

[1] Aourir M, Abouloifa A, Lachkar I, Aouadi C, Giri F, Guerrero JM. Nonlinear control and stability analysis of single stage grid-connected photovoltaic systems. International Journal of Electrical Power & Energy Systems. 2020. ISSN 0142-0615;**115**: 105439. DOI: 10.1016/j.ijepes.2019.105439. Available from: https://www.sciencedirect.com/science/article/pii/S0142061519304260

[2] Aouadi C, Abouloifa A, Aourir M, Lachkar I, Tighazouane B, Boussairi Y. Nonlinear control of double stage three-phase grid-connected photovoltaic systems. IFAC-PapersOnLine. 2022. ISSN 2405-8963;**55**(12):526-531. DOI: 10.1016/j.ifacol.2022.07.365. Available from: https://www.sciencedirect.com/science/article/pii/S2405896322007662. 14th IFAC Workshop on Adaptive and Learning Control Systems ALCOS 2022

[3] Swetha K, Sivachidambaranathan V. A nonlinear controller for neutral point piloted (T-type) multilevel inverter-based three-phase four-wire DSTATCOM. International Transactions on Electrical Energy Systems. 2022; **2022**(1):7899765. DOI: 10.1155/2022/7899765

[4] Boopathi R, Indragandhi V. Solar photovoltaic-interfaced shunt active power filter implementation for power quality enhancement in grid-connected systems. International Journal of Circuit Theory and Applications. 2023;**51**(11): 5305-5323

[5] Aghbalou N, Charki A, Errousso H, Filali Y. Ensemble learning method for forecasting HVAC system demand. In: Elkhattabi EM, Boutahir M, Termentzidis K, Nakamura K, Rahmani A, editors. Advanced Materials for Sustainable Energy and Engineering. Cham, Switzerland: Springer Nature; 2024. pp. 1-9

[6] Kyeremeh F, Fang Z, Liu F, Peprah F. Techno-economic analysis of reactive power management in a solar PV microgrid: A case study of Sunyani to Becheam MV feeder, Ghana. Energy Reports. 2024. ISSN 2352-4847;**11**:83-96. DOI: 10.1016/j.egyr.2023.11.031. Available from: https://www.sciencedirect.com/science/article/pii/S2352484723015561

[7] Rastogi M, Ahmad A, Bhat AH. Performance investigation of two-level reduced-switch D-STATCOM in grid-tied solar-PV array with stepped P&O MPPT algorithm and modified SRF strategy. Journal of King Saud University - Engineering Sciences. 2023. ISSN 1018-3639;**35**(6):393-405. DOI: 10.1016/j.jksues.2021.06.008. Available from: https://www.sciencedirect.com/science/article/pii/S101836392100091X

[8] Varshney G, Mittal U, Pawar A. Power Quality Enhancement Using Passive Power Filters. Vol. 407. Wiley Online Library; 2023. p. 2100442. DOI: 10.1002/masy.202100442. Available from: https://onlinelibrary.wiley.com/doi/abs/10.1002/masy.202100442

[9] Sahoo B, Alhaider MM, Rout PK. Power quality and stability improvement of microgrid through shunt active filter control application: An overview. Renewable Energy Focus. 2023;**44**:139-173

[10] Michalec Ł, Jasiński M, Sikorski T, Leonowicz Z, Jasi Ł, Suresh V. Impact of harmonic currents of nonlinear loads on power quality of a low voltage network–review and case study. Energies. 2021;**14** (12):1-19. DOI: 10.3390/en14123665.

Available from: https://www.mdpi.com/1996-1073/14/12/3665

[11] Aourir M, Abouloifa A, Aouadi C, El Otmani F, Lachkar I, Giri F, et al. Nonlinear control of multicellular single stage grid connected photovoltaic systems with shunt active power filtering capability. IFAC-PapersOnLine. 2020. ISSN 2405-8963;**53**(2): 12853-12858. DOI: 10.1016/j.ifacol.2020.12.2091. Available from: https://www.sciencedirect.com/science/article/pii/S2405896320327427. 21st IFAC World Congress

[12] Ahmed K, H, Finney SJ, Williams BW. Passive filter design for three-phase inverter interfacing in distributed generation. In: Compatibility in Power Electronics. IEEE; 2007. CPE'07. pp. 1-9

[13] Aourir M, Abouloifa A, Aouadi C, El Otmani F, Lachkar I, Noussi K, et al. Hybrid control of flying capacitor converter based shunt active power filter for photovoltaic systems interconnection. IFAC-PapersOnLine. 2022. ISSN 2405-8963;**55**(12):49-54. DOI: 10.1016/j.ifacol.2022.07.287. Available from: https://www.sciencedirect.com/science/article/pii/S2405896322006887. 14th IFAC Workshop on Adaptive and Learning Control Systems ALCOS 2022

[14] Campanhol LBG, da Silva SAO, Azauri AO. A three-phase four-wire grid-connected photovoltaic system using a dual unified power quality conditioner. In: 2015 IEEE 13th Brazilian Power Electronics Conference and 1st Southern Power Electronics Conference (COBEP/SPEC), Fortaleza, Brazil. 2015. pp. 1-6. DOI: 10.1109/COBEP.2015.7420130

[15] Amini B, Rastegar H, Pichan M. An optimized proportional resonant current controller based genetic algorithm for enhancing shunt active power filter performance. International Journal of Electrical Power & Energy Systems. 2024ISSN 0142-0615;**156**:109738. DOI: 10.1016/j.ijepes.2023.109738. Available from: https://www.sciencedirect.com/science/article/pii/S0142061523007950

[16] Annamalaichamy A, David PW, Balachandran PK, Colak I. Performance evaluation of PI and FLC controller for shunt active power filters. Electrical Engineering (Springer Science and Business Media LLC). 2024:1-18

[17] Hekss Z, Abouloifa A, Echalih S, Lachkar I, El Aroudi A, Giri F. Flatness based control of single phase grid connected photovoltaic system associated with a shunt active power filter. IFAC-PapersOnLine. 2022. ISSN 2405-8963;**55**(12):43-48. DOI: 10.1016/j.ifacol.2022.07.286. Available from: https://www.sciencedirect.com/science/article/pii/S2405896322006875. 14th IFAC Workshop on Adaptive and Learning Control Systems ALCOS 2022

[18] Aourir M, Abouloifa A, El Otmani F, Aouadi C, Hamdoun A, Lachkar I. Flatness based control of single phase multicellular shunt active power filter for power quality improvement. In: 2018 Renewable Energies, Power Systems & Green Inclusive Economy (REPS-GIE), Casablanca, Morocco. 2018. pp. 1-6. DOI: 10.1109/REPSGIE.2018.8488853

[19] Aouadi C, Abouloifa A, Lachkar I, Aourir M, Boussairi Y, Hamdoun A. Nonlinear controller design and stability analysis for single-phase grid-connected photovoltaic systems. International Review of Automatic Control (IREACO). 2017;**10**(4):306-315. DOI: 10.15866/ireaco.v10i4.12322

Chapter 7

# Downhole Steam Generation for Heavy Oil and Oil Sands Production

*Kent B. Hytken*

## Abstract

Steam injection lowers the viscosity of heavy oil and oil sands trapped in underground formations. Steam injection is one of the most reliable thermal-enhanced oil recovery methods for heat introduction. Conventional steam applications generate steam from an inefficient high-pressure boiler, which has significant heat losses, degradation in steam quality, and uncertainty in steam distribution in the reservoir while emitting greenhouse gas emissions into the environment. In contrast to the high-pressure steam boiler, a downhole steam generator provides a step-change improvement with immense potential to produce heavy oil below 2500 ft., which can be a transformative thermal-enhanced oil recovery method and pave the way for an exciting advancement.

**Keywords:** downhole steam generation, thermal-enhanced oil recovery, steam-assisted gravity drainage (SAGD), cyclic steam injection, heavy oil extraction

## 1. Introduction

Undoubtedly, thermal-enhanced oil recovery (TEOR) techniques, such as steam injection, offer significant benefits. However, their implementation is not without its share of challenges. These techniques demand substantial upfront capital investments and ongoing operational expenses, which can strain finances. Additionally, the need for specialized equipment, facilities, and energy sources can further complicate these projects. It is crucial to understand these challenges to fully appreciate the need for innovative solutions like downhole steam generation [1]. The urgency for such solutions is clear in the face of these challenges.

Downhole steam generation (DHSG) is a groundbreaking technology emerging as a viable alternative to conventional steam injection for heavy oil recovery. Unlike conventional methods that generate steam at the surface and transport it downhole, DHSG creates steam directly within the reservoir. This innovative approach not only eliminates significant heat losses during transport but also has the potential to reduce environmental impact, offering exciting possibilities for maximizing heavy oil extraction while promoting sustainability in oil recovery [1].

Heavy oil and oil sands are the largest oil reserves in the world, representing two-thirds of the total oil reserves (**Figure 1**). The two main TEOR applications are steam flooding and cyclic steam stimulation (CSS), but they have certain limitations that need further improvement [2].

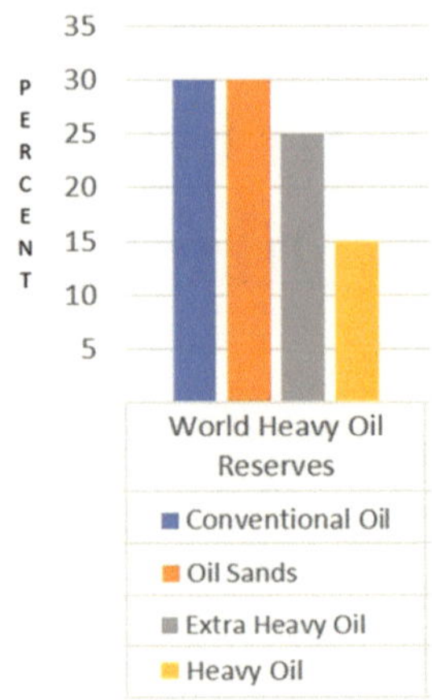


**Figure 1.**
*Percentage of world heavy oil reserves. Source: Future Energy LLC.*

## 2. History of steam generation for thermal-enhanced oil recovery

Steam injection processes recover heavy oil and oil sands. Since the first known application of steam in a Venezuelan oil field in the late 1950s, worldwide heavy oil production has undergone phenomenal growth [1] attributed to three factors:

- Higher and more stable oil prices;
- Lower capital and operating costs;
- Technological advances.

Technological advances, spanning geological, geophysical, horizontal well advances, and production engineering disciplines have helped reduce costs and significantly improved recovery. These advancements are a beacon of hope, promising a brighter future for heavy oil production [1].

It has long been recognized that applying heat to heavy oil reservoirs can substantially improve recovery. Although steam injection is currently the principal thermal recovery method, heat transmission losses associated with the delivery of the steam from the surface generators to the oil-bearing formation are limited to shallow reservoirs above 2000 ft. [1].

The objective of the Department of Energy's Project DEEP STEAM was to develop the technology required to produce heavy oil from deep reservoirs economically. This effort included developing and evaluating thermally efficient delivery systems and a downhole steam generation system. In 1976, the United States Department of Energy (DOE) and Sandia National Laboratories (Sandia) began a downhole steam generator project known as "Project Deep Steam." This project was based on the premise that significant reserves of worldwide heavy oil were beyond the reach of surface-generated steam. The DOE-funded Sandia's $26 million budget over 6 years. In 1982, the DOE and Sandia began a field development test in Long Beach, California, to place two direct contacts, high-pressure combustors in the Wilmington Oil Field. The field test was not successful [3].

Heavy oil reserves are the largest proven oil reserves in the world, representing two-thirds of the total oil reserves. The percentage of over two trillion barrels of deep (below 2500 ft.) worldwide heavy oil, oil sands, and bitumen resources is shown below (**Figure 1**).

### 2.1 Conventional steam injection

Surface steam injection produces heavy oil at shallow depths above 2000 ft. Since the early 1960s, the antiquated high-pressure once-through steam generator (OTSG) has been the primary thermal stimulation technique for heavy oil and oil sands reservoirs [4]. An OTSG generates steam in the oil field and suffers its most significant drawback in heat losses as it flows through the surface piping to the injection well, and the most heat losses are in the wellbore.

The two main conventional TEOR applications are steam flooding and cyclic steam stimulation (CSS), but they have limitations that need further improvement [1]. Conventional steam injection mainly utilizes a high-pressure once-through steam generator (OTSG) boiler for surface for cyclic steam stimulation (CSS), steam flooding, or steam-assisted gravity drainage (SAGD). The OTSG boiler is inefficient and experiences significant heat losses, degradation in steam quality, and uncertainty in steam distribution in the reservoir while emitting greenhouse gas (GHG) emissions into the environment (**Figure 2**).

In contrast to the high-pressure OTSG boiler, a downhole steam generator provides a step-change improvement in capital and operating costs while delivering controlled and predictable high-quality saturated or superheated steam to the reservoir for improved reservoir management, steam distribution, and heavy oil recovery.

### 2.2 Cyclic steam stimulation

Cyclic steam stimulation (CSS), sometimes called steam huff-and-puff or steam soak, is a type of thermal-enhanced oil recovery (TEOR) used explicitly for heavy oil reservoirs. Unlike steam flooding, a continuous process, CSS involves a series of cycles of injecting steam, soaking, and production [1].

**Figure 2.**
*Conventional OTSG boiler. Source: www.en-fabinc.com.*

Three-step process of cyclic steam stimulation (CSS):

- *Steam injection.* Steam is injected into a well for a designated period, causing heat to penetrate deep into the reservoir.

- *Soak period.* The well is then shut in, allowing the injected steam to soak into the formation and transfer heat to the surrounding rock and oil. This heating process reduces the viscosity of the heavy oil, making it flow more easily. The soak period can last anywhere from a few days to a week.

- *Production phase.* The well is switched back to production mode after the soak period. The heated, thinner oil flows more readily toward the wellbore and can be extracted.

This CSS injection, soak, and production cycle is repeated multiple times over the well's lifetime. Some of the oil is recovered with each cycle, but the process becomes less effective as the reservoir cools down over time.

Several critical points about cyclic steam stimulation (CSS):

- *Relatively simple and low-cost.* CSS is more straightforward and cheaper than other enhanced oil recovery methods.

- *Effective for heavy oil*. It is particularly well-suited for extracting viscous heavy oil that would not flow easily under normal conditions.

- *Limited recovery and environmental impact.* While effectively boosting initial production rates, CSS has a lower overall oil recovery than other TEOR methods. Additionally, the energy used to generate steam can contribute to greenhouse gas emissions.

Cyclic steam stimulation (CSS) is a widely used technique for recovering heavy oil, but it has limitations regarding total yield and environmental impact. CSS heats the oil in the reservoir to recover 20–25% of the original oil in place (OOIP) (**Figure 3**).

### 2.3 Steam flooding

Steam flooding is an efficient method for recovering significant amounts of oil from reservoirs that would not be productive with conventional techniques [1].

The attributes of steam flooding:

- Steam is injected into underground reservoirs through designated injection wells.

- The intense heat from the steam thins the heavy oil by reducing its viscosity, making it flow more easily toward production wells.

- Some of the steam condenses into hot water as it travels through the reservoir, further aiding in mobilizing the oil.

- In some cases, the heat can also vaporize lighter oil components, creating a solvent that helps extract even more oil.

*DOI: http://dx.doi.org/10.5772/intechopen.1005737*

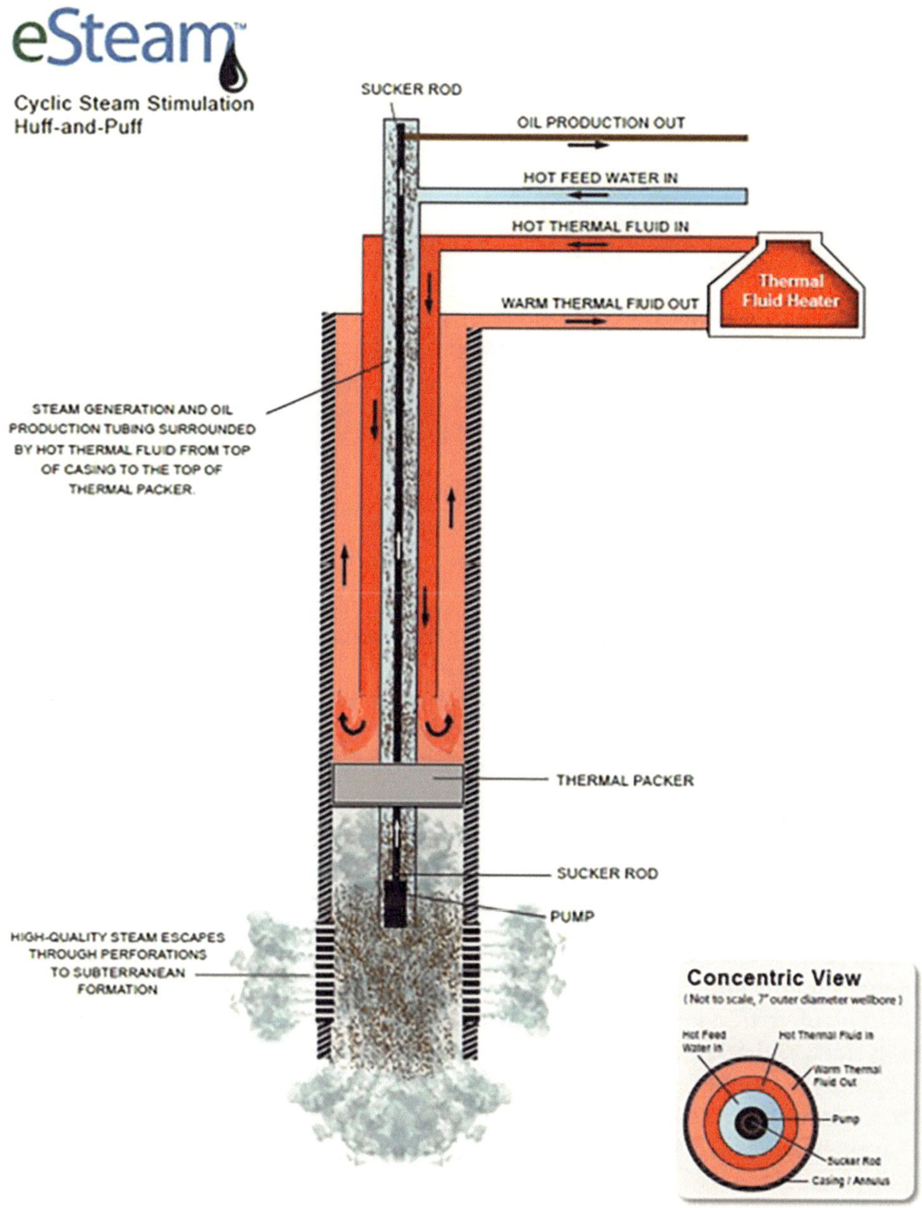


**Figure 3.**
*Cyclic steam stimulation (CSS). Source: www.esteamoil.com.*

It is essential to consider some limitations:

- Steam flooding is an energy-intensive process, often requiring burning fossil fuels to generate steam, which can contribute to greenhouse gas emissions.

- Steam flooding is not suitable for all heavy oil reservoirs. Factors like reservoir thickness, porosity, and water content can affect effectiveness.

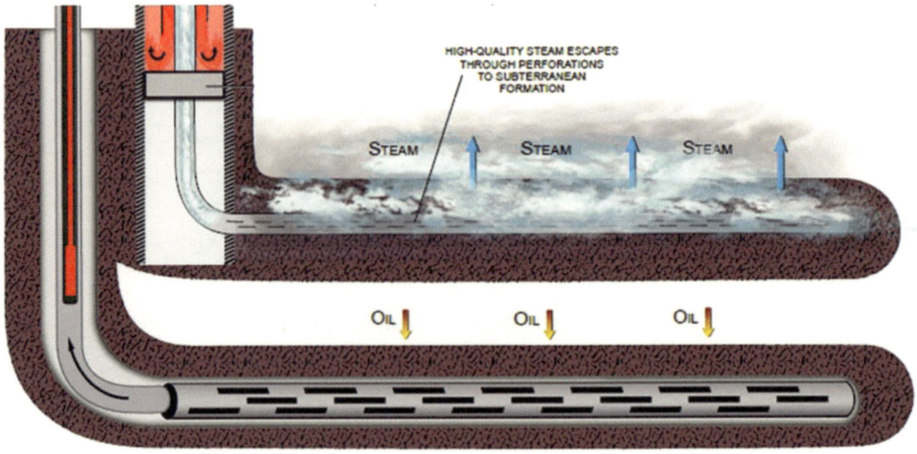


**Figure 4.**
*Steam-assisted gravity drainage (SAGD). Source: www.esteamoil.com.*

Steam flooding is valuable for extracting heavy oil but requires careful planning and execution to balance its benefits with environmental considerations. A form of steam flooding that has become popular in Alberta, Canada's oil sands is steam-assisted gravity drainage (SAGD), in which two horizontal wells are at least 10 meters above the other well [5]. The steam is injected into the upper wellbore to heat the heavy oil and reduce its viscosity, causing the heated heavy oil to drain into the lower wellbore and be pumped out of the producing horizontal well. Steam rises in an up-and-out pattern, forming a steam chamber. This chamber mobilizes the oil drained by gravity and captured by the lower horizontal production well. Once the chamber reaches the top of the reservoir formation, lateral growth takes over, and output reaches a maximum plateau, which slowly declines. Two primary mechanisms exist for SAGD: ceiling drainage and slope drainage. Ceiling drainage occurs when the steam chamber rises and expands, causing the oil to flow in a countercurrent pattern. Ceiling drainage plays a dominant role in the SAGD process and is a function of vertical permeability. Slope drainage occurs when the steam chamber expands sideways horizontally and plays a more dominant role in the later stages of the SAGD process [6, 7]. SAGD is a continuous steam injection method (**Figure 4**).

Dr. Roger Butler was an engineer at Imperial Oil Canada from 1955 to 1982, who invented the SAGD process in the 1970s. Dr. Butler developed the concept of using horizontal pairs of wells to develop bitumen that is considered too deep for mining [8]. In 1983, Dr. Butler became director of technical programs for the Alberta Oil Sands Technology and Research Authority.

As of 2023, Canada's oil sands industry, Western Canada, and offshore petroleum facilities near Newfoundland and Labrador continued to increase production. Production is projected to increase by an estimated 10% in 2024, representing a potential record high at the end of the year of approximately 5.3 million barrels per day. The surge in production is mainly due to growth in Alberta's oil sands [8].

### 2.4 Downhole steam generation

Downhole steam generation (DHSG) is an emerging technology in heavy oil recovery that aims to address the limitations of traditional methods. Instead of generating steam at the surface and injecting it downhole, DHSG produces steam directly within the reservoir. DHSG offers potential benefits like eliminating heat loss during transport from the surface. Steam generated at the surface is currently the most common methodology for thermal stimulation of heavy oil and oil sands at shallow depths above 2000 ft. [1, 9]. A downhole steam generator can unlock deeper deposits below 2500 ft., where conventional surface-generated steam is impossible.

The introduction of eSteam's™ downhole steam generator (eSteam™) presents significant benefits for steam injection in shallow and deep reservoirs. eSteam™ prevents heat losses and steam quality degradation, decreases environmental effects, and economically enhances heavy oil and oil sands recovery with significantly reduced capital and operating costs compared to conventional OTSG boilers [10].

eSteam™ is suitable for recovering shallow and deeper heavy oil deposits worldwide below 2500 ft. eSteam™ is a novel and transformative technology that can unlock deep heavy oil resources in environmentally challenged locales such as offshore or under arctic or permafrost environments [10].

eSteam's™ novel downhole design improves the delivery of thermal energy in vertical, horizontal, and lateral wellbores in geological subterranean formations that achieves high-pressure, high-quality saturated or superheated steam that increases the efficiency of recovering viscous hydrocarbons from subterranean formations. eSteam™ can also achieve steam fracking of low-permeability subterranean formations such as diatomite, tight oil, and shale oil [10].

eSteam's™ downhole design includes concentric tubing strings in the wellbore and a thermal packer. A non-toxic hot, heat transfer fluid circulates in a closed-loop concentric tubing configuration coupled to the outermost tubing string, and the hot, heat transfer fluid returns to the surface to be reheated in the low-pressure thermal fluid heater and recirculates downhole. The hot feedwater continually circulates through the hot innermost tubing string from the heat transfer fluid that acts as the heat exchanger and is concentric to the outermost tubing string, eliminating heat loss (**Figures 5** and **6**) [9].

eSteam's™ high-quality saturated or superheated steam generation can facilitate steam fracking of tight oil or low-permeability subterranean formations, which may result in the propagation of fractures in the formation of the rock layer. Steam fracking is a technique that fractures the rock layer to enhance hydrocarbon recovery. Steam fracking points to the vast amount of low-volume produced viscous heavy oil

**Figure 5.**
*8 MMBtu/hr. thermal fluid heater. Source: www.esteamoil.com.*

**Figure 6.**
*60 MMBtu/hr. thermal fluid heater. Source: www.cai3.com.*

in low-permeability diatomite, shale oil, and tight oil formations. Steam fracking eliminates potential environmental impacts, including groundwater contamination, air quality risks, the migration of gases, and hydraulic fracturing chemicals that can lead to adverse short-term and long-lasting health effects [9].

Combining eSteam's™ high-quality saturated or superheated steam with other methods will also enhance the recovery of the deposits. For example, different techniques may include the addition of an alkali, such as sodium carbonate, which makes the in-situ generation of a surfactant possible. In addition to reducing interfacial tension to ultra-low values, surfactants and alkalis alter wettability to enhance oil recovery. An alkaline surfactant process enhances spontaneous imbibition in fractured, oil-wet, carbonate formations [9]. Mobility control is essential for steam-surfactant-enhanced oil recovery (EOR) to improve the sweep efficiency into fractured reservoirs. Delivering a catalyst in the formation can enhance heavy oil recovery by upgrading the oil, achieving a higher API gravity oil by increasing the saturate and aromatic components, and reducing the resin and asphaltene components.

The term "aquathermolysis" describes the chemical interaction of high-temperature, high-pressure water with the reactive components of heavy oil and bitumen. In aquathermolysis, the metal species added to steam interact with organic sulfur compounds. In a huff-n-puff operation, the aquathermolysis catalyst will reduce the oil viscosity by more than 60% and increase the oil production for the steam cycle. The metal species in the catalyst for this improvement contained VO2+, Ni2+, and Fe3+. The vanadyl sulfate and nickel sulfate are the catalysts for the aquathermolysis of heavy oils, and ferric sulfate is the catalyst for the water-gas shift reaction. The aquathermolysis process produces hydrogen sulfide ($H_2S$) because of the desulfurization of heavy crude. Gaseous $H_2S$ promotes the water-gas shift reaction through the intermediate formation of carbonyl sulfide (COS). At the same time, $H_2S$ will react with the metal ions and produce metal sulfides. It is well known that metal sulfides are valuable catalysts for hydro-desulfurizing heavy oil. The analysis found that all transition metal species can accelerate the decomposition of the sulfur compounds regardless of whether the sulfur was in an aromatic or an aliphatic environment. Among all the transition metal species, VO2+ and Ni2+ are the most effective for the aquathermolysis of heavy oil [9].

It is determined that the catalyst can significantly alter the composition of heavy oil. Therefore, upgrading heavy oil in the reservoir achieves a higher API gravity oil by removing the asphaltenes. Activating the silica column and obtaining the saturate and aromatic fractions by elution with hexane results in an oil composition change. After the catalytic treatment, the oil has lighter saturate and aromatic components and less resin and asphaltene, improving the overall efficiency of aquathermolysis in transforming heavy crude into a lighter, more valuable product [11].

The catalyst enables the conversion of the hydrocarbons into aromatics during aromatization. At the edge of the condensed aromatic core in resin and asphaltene molecules, standard iso-alkyl side chains broke off from the condensed aromatic and then converted into alkyl hydrocarbons. The alkyl chain, which links two condensed aromatics in large molecular structures of resin and asphaltene, decreased, and the aromatics increased [12].

In the process, heavy oil undergoes aquathermolysis during steam injection, and the catalyst injected with steam can accelerate the reaction, resulting in a reduced viscosity and a changed composition of the produced oil that results in higher API gravity lighter oil. The metallic ions can interact with water with a catalyst added to the steam. The proton from the complex molecule can attack the sulfur atom, and the hydroxide ion can attack the carbon atom. This results in the electronic cloud excursion and leads to a further decrease in bond energy. Because of this, the C-S bond will break in the aquathermolysis process, resulting in a low amount of sulfur and heavy components such as resin and asphaltene [13].

eSteam's™ transformative downhole steam generator prevents wellbore heat losses from degrading the steam quality, and the reservoir receives the highest quality saturated or superheated steam, pressure, and temperature. eSteam's™ unique, innovative methodology attains enhanced heat efficiency for significantly less energy, resulting in faster oil production for a lower cost per pound of steam [10].

The cost of steam per pound (lb.) depends on several factors:

- *Fuel source.* The fuel used to generate the steam (natural gas, diesel, coal, etc.) significantly impacts the cost.

- *Boiler efficiency.* How efficiently the boiler converts fuel into steam affects the overall cost per pound.

- *Production scale.* Large-scale industrial facilities might have different costs than smaller operations.

- *Location.* Energy costs can vary depending on geographic location.

eSteam's™ benefits compared to a conventional OTSG boiler:

- Increased energy efficiency from 45% to ~58%.

- Precise temperature control improves heat efficiency.

- Precise downhole steam quality injected into the reservoir.
- Predictable delivery of a known amount of heat to the formation.
- Steam eliminates heat loss in the wellbore.
- Delivers the entire British Thermal Unit (BTU) content of the fuel.
- Improves steam distribution in the reservoir.
- No moving parts in the wellbore.
- Steam delivered to depths of 700 ft. to ~8000 ft.
- Tolerates brackish feedwater.
- In situ oil upgrading of heavy oil and oil sands.
- Steam fracking of low-permeability diatomite, shale oil, and tight oil.
- Low-pressure thermal fluid heater is scalable up to 120 MMBtu/hr.

eSteam™ achieves better economics than a conventional OTSG boiler:

- Projected ~45% more volume of steam.
- Lower steam-oil ratio (SOR) 1.75 to 2:25.
- ~$19.00/Bbl all-in production cost.
- ~50% lower cost per barrel of steam.
- ~25–30% lower operating expenses (OpEx).
- ~50% reduction in capital expenditure (CapEx).

eSteam™ eclipses the conventional OTSG boiler with a more efficient steam system that generates high-quality saturated or superheated steam in the formation and achieves better steam distribution in the reservoir and a lower steam-oil ratio (SOR) than conventional surface steam generation. An important fact is that surface steam injected in a SAGD well, assuming the steam descends 1000 ft. down the vertical well and then 1000 ft. in the horizontal leg, will condense to 100% hot water about 1/3–1/2 of the way in the horizontal well, resulting in a hot water flood for the balance of the horizontal well. eSteam™ generates steam right above the radius of the horizontal leg, and the pressure carries the steam into the horizontal leg, reaching the toe before condensing to 100% hot water. The feedwater does not have to be fresh water used in an OTSG boiler since the heat transfer occurs downhole [9, 10].

eSteam™ is suitable for recovering heavy oil deposits worldwide that have experienced low recovery efficiencies through primary and secondary (waterflood) production methods. These deeper heavy oil resources typically require thermal stimulation to reduce viscosity in situ for improved production volumes [1]. eSteam™ operates with limited maintenance and downtime for extended periods since there are no moving equipment parts downhole in the wellbore [10]. eSteam™ has overcome the technological constraints of deep heavy oil production.

eSteam™ is not an electrical or combustion technology that eliminates a downhole combustion system. eSteam™ has mitigated the technical challenges, including downhole concentric tubing, which provides operational safety, significantly reduces wellbore heat losses, and creates optimum steam quality, pressure, and temperature in the reservoir, providing better steam distribution and improving sweep efficiency [9, 10].

eSteam's™ proprietary hydrogen boiler eliminates all NOx and $CO_2$ emissions through closed-loop combustion. It is a breakthrough high-temperature boiler that generates heat, with water as the only by-product. No air permit is required. It is 20% more efficient than conventional steam boilers, with 95% thermal efficiency (**Figures 7–10**).

eSteam's™ key characteristics compared to conventional steam applications:

- *Variable depth capability.* Deploying eSteam™ to a deep pay zone results in significantly higher operating efficiencies with zero thermal loss of the steam in the wellbore and controlled injection down to about ~8000 ft. [10].

- *Capital expenditure (CAPEX).* eSteam™ is ~65% less capital cost than conventional steam applications [10].

- *Operating expenses (OPEX).* eSteam™ has 25–30% less operating expenses than conventional steam boilers, for example, OTSG [10].

- *Lower steam-oil ratio.* eSteam's™ high-quality saturated or superheated steam in the formation achieves better steam distribution in the reservoir and a lower steam-oil ratio (SOR) than conventional surface steam generation [10].

**Figure 7.**
*Zero-emissions hydrogen boiler. Source: www.hydrogentechnologiesllc.com.*

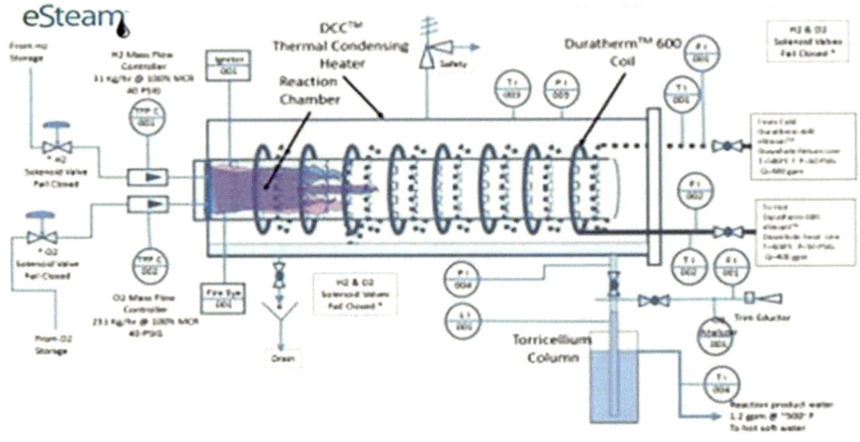


**Figure 8.**
*Zero-emissions hydrogen heater diagram. Source: www.esteamoil.com and www.hydrogentechnologiesllc.com.*

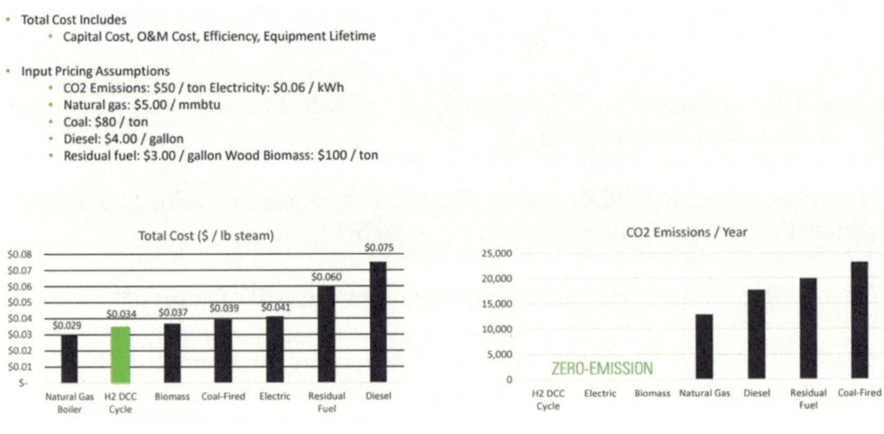


**Figure 9.**
*Cost-competitive advantages. Source: www.hydrogentechnologiesllc.com.*

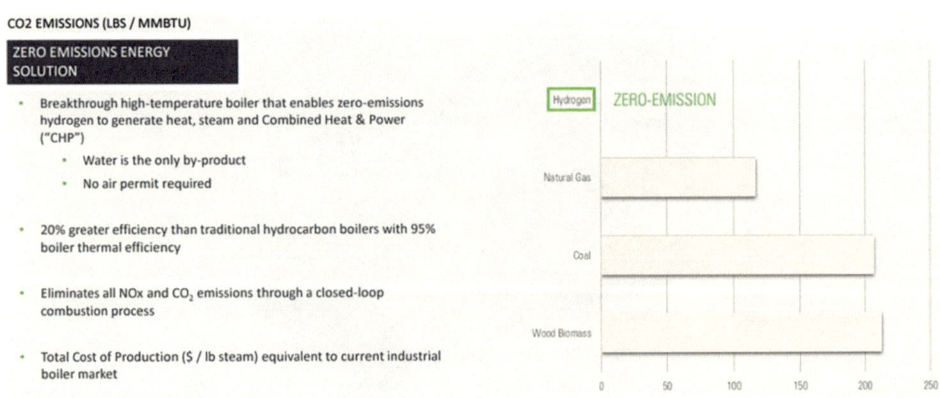


**Figure 10.**
*$CO_2$ emissions (Lbs./MMBtu). Source: www.hydrogentechnologiesllc.com.*

- *Water savings.* eSteam's™ ability to deploy at variable depths requires less water usage than conventional steam boilers, for example, OTSG [10].

- *Energy savings.* eSteam™ delivers the entire BTU fuel content compared to the inefficient OTSG boiler, including the higher heating value and avoiding boiler losses, surface steam distribution, and wellbore losses [10].

- *Off-the-shelf proven equipment.* Off-the-shelf proven equipment (also referred to as COTS—Commercial Off-The-Shelf) refers to readily available, pre-made equipment that has a proven track record of successful use in similar applications [10].

Critical characteristics of off-the-shelf proven equipment:

- *Commercially available.* It can be purchased from a supplier without extensive customization.

- *Standardized design.* It follows a standard and mass-produced design, making it readily available and often more affordable.

- *Proven performance.* It has been successfully used in similar applications, demonstrating its reliability and effectiveness.

- *Lower development costs.* Compared to custom-made equipment, it requires less engineering and design work, leading to lower upfront costs.

Advantages of using off-the-shelf proven equipment:

- *Faster implementation.* Since it is readily available, it can be deployed more quickly than custom-made equipment.

- *Reduced costs.* Lower development costs and potentially lower upfront purchase price than custom options.

- *Lower risk.* A proven track record in similar applications reduces the risk of unexpected performance issues.

eSteam's™ proprietary software operates a programmable logic controller (PLC) that controls the above-ground equipment, providing broad variable control of temperatures, pressures, and steam injection rates into the formation. This ensures improved heat efficiency, precise steam quality injected, and predictable delivery of a known amount of heat to the formation for improved steam distribution in the reservoir (**Figure 11**).

The market opportunity for thermal-enhanced oil recovery is quite substantial. Sales of antiquated conventional OTSG boilers were about $1 billion in 2022. The global thermal-enhanced oil recovery (EOR) market for boilers and other equipment is expected to surpass US$30 billion in 2034 [14].

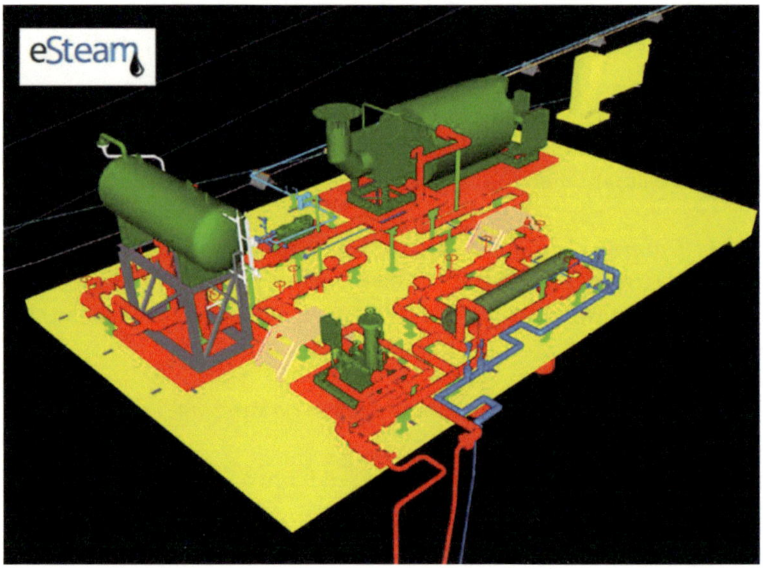


**Figure 11.**
*eSteam™ equipment on the pad site. Source: www.esteamoil.com.*

## 3. Conclusion

Heavy oil is highly viscous, and the predominant method of heating heavy oil involves injecting steam into the reservoir. The problem when injecting steam downhole from above-ground steam generators is that the steam loses its quality and thermal energy as depth increases. The steam quality dissipates in heavy oil formations below 2000 ft., impairing oil recovery. The ability of eSteam™ to inject directly at the pay zone without heat loss eliminates the additional energy cost needed to inject from the surface, allowing the operating/lift costs to remain constant below 2500 ft.

Downhole steam generation is an emerging technology alternative to conventional steam applications. Among the processes applied to recover low-gravity viscous crude oil, thermal methods are the most successful because they increase the oil production rate and ultimate recovery. Thermal-enhanced oil recovery (TEOR), such as steam injection, involves applying heat to the reservoir to improve oil mobility and increase heavy oil recovery. Implementing thermal EOR techniques involves substantial upfront capital investments and ongoing operational expenses. The need for specialized equipment, facilities, and energy sources can make these projects financially challenging.

The two basic injection techniques, cyclic steam stimulation (CSS) and steam flooding have gained momentum in steam-assisted gravity drainage (SAGD) wells. CSS (or soak, cyclic, and huff-n-puff) is a single-well operation. Oil viscosity is the chief variable because heavy oil recovery depends primarily on reducing viscosity. The ultimate heavy oil recovery is low. Steam flooding requires a set of injector wells for steam injection and producer wells for oil production. Steam distillation may be necessary to increase total recovery efficiencies. Heavy oil production is a continuous operation but slow oil production. However, the ultimate oil recovery is high. SAGD produces heavy crude oil and bitumen from horizontal wells in the reservoir. A conventional once-through steam generator (OTSG) boiler generates steam in the oil field and suffers its most significant heat losses as it flows through the surface piping to the injection well. The most heat losses are in the wellbore, which cannot deliver steam in deep wells below 2000 ft. Alternatively, eSteam's™ primary advantage is that it eliminates surface and wellbore heat losses and achieves high-quality saturated or superheated steam above and below 2500 ft.

## Conflict of interest

The author declares there is no conflict of interest.

## Nomenclature

1. *Enhanced oil recovery (EOR) methods.*

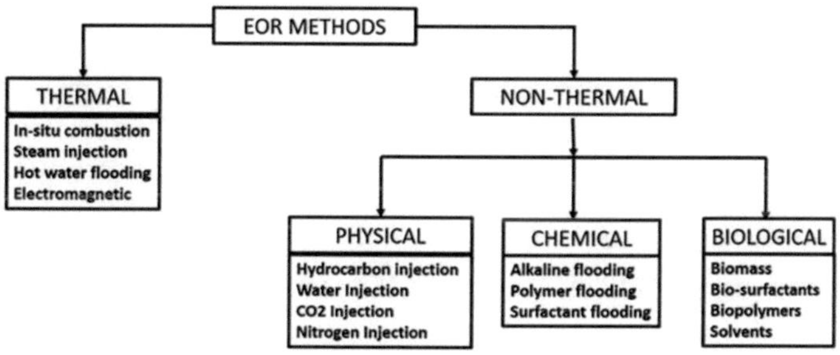


2. *Thermal-enhanced oil recovery (TEOR)*. TEOR is used in the oil industry to extract heavy oil that does not flow easily under normal conditions. By injecting heat into the reservoir, TEOR reduces the oil's viscosity, making it thinner and more accessible to pump to the surface.

3. *Steam flooding*. Steam flooding, or continuous steam injection or steam drive, is a specific type of TEOR used to extract heavy oil reserves.

4. *Cyclic steam stimulation (CSS)*. CSS, sometimes called steam huff-and-puff or steam soak, is a type of thermal-enhanced oil recovery (TEOR) used explicitly for heavy oil reservoirs. Unlike steam flooding, a continuous process, CSS involves a series of cycles of injecting steam, soaking, and then production.

5. *Cyclic huff-n-puff steam injection, also known as CSS*. Huff-n-puff is a method for extracting heavy oil by repeatedly heating the reservoir with steam. In each cycle, high-pressure steam is injected into a well; the well is shut for a soaking period to allow the heat to thin the oil. Finally, the well is switched back to production to recover the mobilized oil.

6. *Steam-assisted gravity drainage (SAGD)*. SAGD is an innovative technique for extracting heavy oil reserves that utilizes a pair of horizontal wells drilled into the reservoir. High-pressure steam is continuously injected into the upper well, creating a heated chamber that thins the heavy oil and reduces viscosity. Gravity then takes over, causing the mobilized oil to drain downwards into the lower production well, which is pumped to the surface.

7. *Once-through steam generator (OTSG)*. An OTSG is used in heavy oil production and can generate clean, high-pressure steam by continuously heating treated fresh feedwater in a single pass-through internal U-tubing. This method is used for steam flooding or cyclic steam stimulation.

8. *Downhole steam generation (DHSG)*. A DHSG is an emerging technology in heavy oil recovery that addresses the limitations of conventional steam applications. Instead of generating steam at the surface and injecting it downhole, DHSG produces steam directly within the reservoir. DHSG eliminates heat loss during transport from the surface.

9. *Steam fracking*. Steam fracking is an emerging technique for heavy oil reservoirs that involves injecting high-pressure steam in low-permeability reservoirs to fracture the formation and heat heavy oil simultaneously. This reduces the oil's viscosity and causes it to flow more easily toward production wells.

10. *Thermal fluid heater*. A thermal fluid heater is used to heat a process or system indirectly. It works by circulating a heat transfer fluid, thermal oil, through a closed-loop system. The heater heats this thermal fluid, which then travels to heat exchangers, transferring its thermal energy to the desired application. Thermal fluid heaters offer precise temperature control, efficient heat transfer, and safer operation than heating with flames or combustion gases.

11. *Saturated steam*. Saturated steam refers to steam that exists at the boiling point of water for a given pressure. This means the steam contains the maximum amount of moisture it can hold at that temperature and pressure. It is effective for heating the reservoir and reducing heavy oil viscosity. Still, compared to superheated steam, it can condense more efficiently as it travels through the cooler reservoir, potentially reducing its heating efficiency.

12. *Superheated steam*. Unlike saturated steam in heavy oil production, superheated steam has been heated beyond its boiling point for a given pressure. This extra heat creates steam with lower moisture content and energy per unit volume. For heavy oil recovery, superheated steam offers advantages because it contains less moisture and is less prone to condensation during travel through the cooler reservoir, delivering more heat deeper into the formation.

13. *Aquathermolysis*. Aquathermolysis is a promising technique for upgrading heavy oil in the reservoir. It utilizes combined effects of water and heat (steam injection) to break down the molecule's rheology in heavy oil. Catalysts are added to enhance these reactions further, improving the overall efficiency of aquathermolysis in transforming heavy crude into a lighter, more valuable product.

14. *Aquathermolysis catalyst*. An aquathermolysis catalyst is a unique substance injected with steam during the aquathermolysis process. These catalysts, often composed of transition metals, act like microscopic accelerators for the chemical reactions that break down heavy oil molecules. This translates to lighter, more valuable crude oil with lower viscosity and a higher flow rate, ultimately enhancing the economic viability of extracting heavy oil resources.

## Author details

Kent B. Hytken
Future Energy LLC, Houston, Texas, United States

*Address all correspondence to: kent@futureenergyllc.com

## References

[1] Vishnumolakala N, Zhang J, Ismail NB. A comprehensive review of enhanced oil recovery projects in Canada and recommendations for planning successful future TEOR projects. In: Paper Presented at the SPE Canada Heavy Oil Conference, September 28–October 2, 2020

[2] Kornienko V, Avtonomov P. Application of Neutron Activation Analysis for Heavy Oil Production Control. Vol. 195. Amsterdam, Netherlands: Elsevier; 3 July 2015. pp. 2451-2456

[3] Moreno JB, Aeschliman DP, Clay RG, Donaldson AB. Project DEEP STEAM 2nd and 3rd Quarter Reports. Albuquerque, NM (USA): Sandia National Labs; 1981

[4] Gwak K-W, Bae W. A review of steam generation for in-situ oil sands projects. Geosystem Engineering. 2010;**13**(3):111-118

[5] Gates ID, Chakrabarty N. Optimization of steam-assisted gravity drainage (SAGD) in ideal McMurray reservoir. Journal of Canadian Petroleum Technology. January 2005;**45**(9)

[6] Mobeen Fatemi S. Injection well-producer well combinations in application of toe-to-heel steam flood. In: Proceedings of SPE EUROPEC Annual Conference and Exhibition. Vienna, Austria; May 2011

[7] Mobeen Fatemi S. Stimulation study of steam-assisted gravity drainage (SAGD) in fractured systems. Oil and Gas Science and Technology. July 2009;**64**(4):477-487

[8] Keyser T. Roger Butler: Father of SAGD. In: Paper Published in Research Gate, March 2005

[9] Hytken KB. "Thermal energy delivery and oil production arrangements and methods thereof" was granted patents by the United States Patent and Trademark Office, US Patent 10,669,828 issued on June 2, 2020, US Patent 11,162,343 issued on November 2, 2021, US Patent 11,788,393 issued on October 17, 2023

[10] Hytken KB. "Methods and systems for enhanced delivery of thermal energy for horizontal wellbores" was granted a patent by the United States Patent and Trademark Office, US Patent 9,464,514 issued on October 11, 2016

[11] Schacht P, Torres-Mancera P, Ancheyta J. Catalytic In-Situ Upgrading of Heavy and Extra-Heavy Crude Oils. Chapter 5. Wiley Online Library; 5 June 2023

[12] Fan H-F, Yongjian L, Liying Z, Xiaofei Z. The study on composition changes of heavy oils during steam stimulation processes. Energy and Fuels. 2002;**81**(13):1733-1738

[13] Fan H-F, Liu Y-J, Zhong L-G. Studies on the synergetic effects of mineral and steam on the composition changes of heavy oils. Energy and Fuels. 2001;**15**(6):1475-1479

[14] Fortune Business Insights Pvt. Ltd. Enhanced Oil Recovery Market Technology (Thermal Injection, Gas Injection, and Chemical Injection). United Kingdom; 1 July 2024

Chapter 8

# The Role of Machine Learning Methods for Renewable Energy Forecasting

*Övgü Ceyda Yelgel and Celal Yelgel*

## Abstract

Across the world, countries are placing greater emphasis on transitioning to cleaner energy sources, while also becoming increasingly concerned about the worsening climate crisis. With the cost-effectiveness and eco friendly nature of renewable energy (RE) sources, there has been a considerable amount of interest. Nevertheless, the unpredictable nature of RE sources presents significant challenges to the security and stability of power grids, adding complexity to the operation and scheduling of power systems. Consequently, the widespread adoption of RE applications becomes more challenging. Accurately forecasting the efficiency of RE is essential for effective system management and operation. By improving the accuracy of these forecasts, we can minimise risks and enhance the stability and reliability of the network. Machine learning (ML) has the potential to greatly assist in achieving the future objectives of RE by comprehending complex correlations within data and providing accurate predictions. This review offers valuable insights into the prediction of RE generation using ML techniques. It explores a wide range of RE sources, such as solar, wind, hydroelectric, geothermal, biomass, and marine-based energies. In addition, the assessment offers a detailed analysis of the latest research findings, along with comprehensive information on performance metrics and ML techniques utilised in RE forecasting.

**Keywords:** renewable energy, machine learning, renewable energy forecasting, artificial intelligence, energy efficiency

## 1. Introduction

Energy is an essential component for fulfilling fundamental requirements and maintaining life. Due to technical advancements, industrialisation, and the fast growth of the global population, there is a growing need for energy. Energy, defined as the capacity to do labour, exists in several forms including potential, kinetic, heat, electrical, light, sound, chemical, and nuclear. It is possible to convert energy from one form to another. Energy resources are often categorised as either renewable or non-renewable based on their use, as represented in **Figure 1**. Currently, countries mostly rely on fossil fuels, including coal, oil, and natural gas, to provide the majority

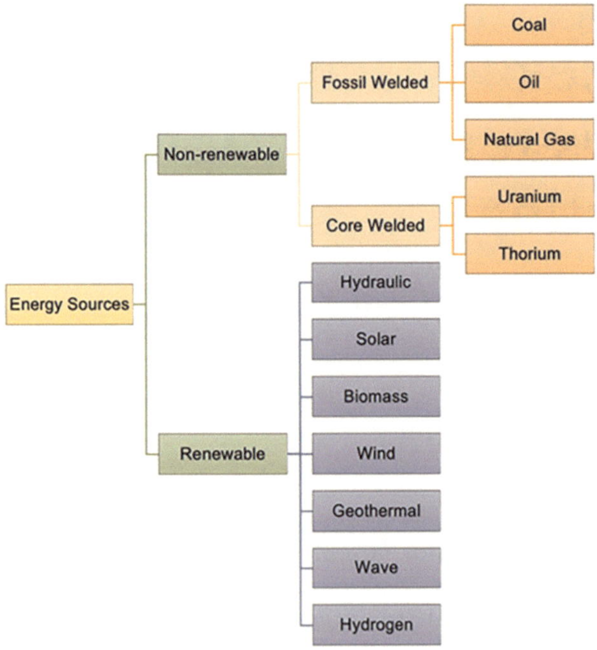


**Figure 1.**
*Energy resources are generally classified as renewable and non-renewable according to their use.*

of their energy requirements. Fossil fuels, which contribute to global warming, also result in air pollution, the development of acid rain, the loss of the ozone layer, and the destruction of forests [1]. The fast advancement of global industrialisation has led to the excessive use of fossil fuels, which not only hastens the depletion of fossil fuel sources but also contributes to global climate change. Consequently, the worldwide utilisation of renewable energy sources is steadily rising. Renewable energies are pristine and limitless energy sources that exhibit enough diversity and abundance, making them suitable for widespread utilisation throughout the globe [2]. As a result of the exhaustion of fossil fuel reserves and the associated environmental harm, nations have begun to shift towards sustainable energy alternatives. Furthermore, it is feasible to satisfy the escalating energy need only by the utilisation of sustainable and environmentally friendly renewable energy sources [3]. Moreover, the implementation and use of sustainable energy systems may serve as a remedy for current energy issues, such as enhancing the dependability of energy supply and resolving localised energy deficits. The benefits of renewable energy sources might be enumerated as follows: As natural resources, they are limitless, decrease nations' reliance on foreign energy, have no impact on global warming or climate change, do not contribute to environmental contamination, emit no harmful emissions, and are sustainable and environmentally friendly sources of energy. Conversely, the substantial upfront expenses associated with constructing power plants and their susceptibility to diurnal and seasonal variations, as well as weather conditions, might be seen as drawbacks of renewable energy sources.

**Figure 2** displays the top ten nations expected to increase their renewable energy capacity by almost 80% from 2021 to 2026. Based on this assessment, China is responsible for around 43% of all renewable energy sources, with the USA, India, and

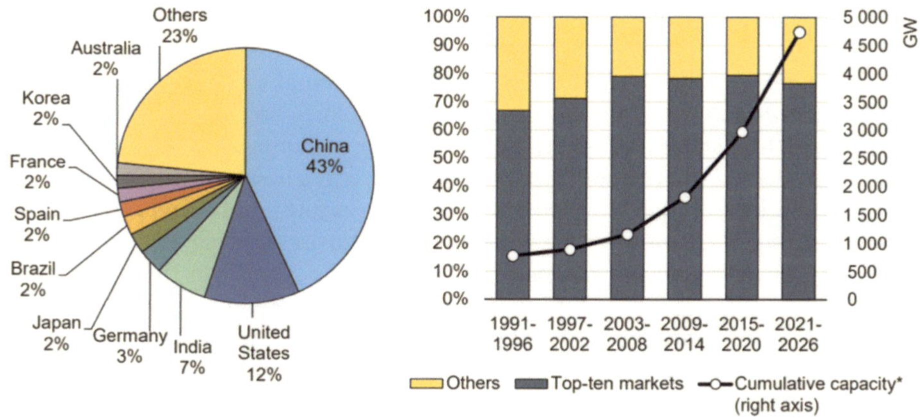


**Figure 2.**
*Share of top ten countries in total installed renewable energy capacity, historical and main situation estimate, 1991–2026 [4].*

Germany following in that order. According to **Figure 2**, the global renewable energy capacity is projected to increase from around 780 GW in the early 1990s to an estimated 4790 GW at now. Furthermore, the top 10 emerging nation markets have made almost equal contributions to the growth of renewable energy capacity from 1991 to 2026. Nevertheless, the market share of nations outside the top ten, despite their significant economic and resource potential, has remained constant since the early 2000s. This trend is projected to persist throughout the projection period.

In the conventional power generating process, the creation of energy relies on the energy requirements of consumers, and the stability of the electrical grid is contingent upon maintaining a balance between energy supply and demand. Insufficient energy supply leads to an imbalance in the electrical system, resulting in power quality issues or blackouts in some areas. When there is an excess of supply compared to demand, energy is wasted, resulting in the incurrence of high unneeded expenditures owing to inefficiency. Ensuring optimal energy production and timing is vital for the efficient functioning of the power system and maximising economic advantages. To preserve this stability, much study has been dedicated to energy supply and demand forecasting, which aims to accurately predict the quantity of energy that will be needed. This prediction will help to limit the surplus of energy. In order to effectively manage an electrical system that relies heavily on renewable energy sources, it is essential to accurately predict the demand for power in the short-term, medium-term, and long-term. This would enhance the development of well-informed energy strategies, for instance, by aiding in the identification of crucial characteristics such as optimal reserve levels and storage needs. However, due to the influence of several unpredictable environmental elements, it is essential to assess the anticipated energy output from renewable energy facilities. This requires the assessment of several environmental variables (such as wind velocity, wind orientation, temperature, pressure, humidity, and solar radiation) within the vicinity of the power plant. Energy forecasting technology is essential for the advancement, administration, and decision-making processes of energy systems.

Forecast models in renewable energy generation may be categorised into three primary groups: physical, statistical, and machine learning. Physical approaches are used as supplementary inputs in statistical models to estimate long-term renewable energy supplies. These methods utilise various physical variables, including air pressure, obstacles, temperature, and topography. According to Du et al. [5], short-term predictions may find long-term renewable energy sources to be inadequate. Statistical models that use historical data on environmental variables impacting renewable energy sources are often favoured, particularly for making predictions in the short-term. Statistical models are inadequate in predicting the nonlinear characteristics of renewable energy sources. Hence, researchers have devised machine learning techniques capable of capturing nonlinear features and delivering more reliable forecasts [6]. Machine learning is a field that focuses on using data to construct prediction techniques and algorithms, first described by Samuel in 1967. The application of machine learning techniques to improve the management of renewable energy production and consumption has yielded positive outcomes, as observed in various studies [6–15]. The key applications of machine learning methods in the field of renewable energy include determining the optimal location, size, and design of renewable energy power plants, as well as optimising the production and consumption of electrical energy. These criteria are contingent upon several circumstances, including proximity to densely populated areas, local climate changes, land availability and prices, logistical considerations, and other relevant infrastructure.

This article aims to comprehensively investigate machine learning models for predicting future renewable energy resources such as solar, wind, hydroelectric, geothermal, biomass, and marine-based energies. The study focuses on renewable energy resource categories, machine learning models, data preprocessing techniques, and machine learning. The text elucidates the process of selecting parameters and evaluating the performance of the models. Various research in the literature has shown the use of diverse machine learning algorithms for forecasting renewable energy. Furthermore, hybrid machine learning models have been specifically developed to enhance the precision of forecasting renewable energy. Different time durations, such as minutes, hours, days, weeks, and years, are used to estimate renewable energy for various forecasting objectives. The evaluation of machine learning models in renewable energy projections and the optimisation of their integration into the grid are often based on forecast accuracy and efficiency. The arrangement of our paper may be succinctly summarised, as seen in **Figure 3**. Firstly, we will provide in-depth

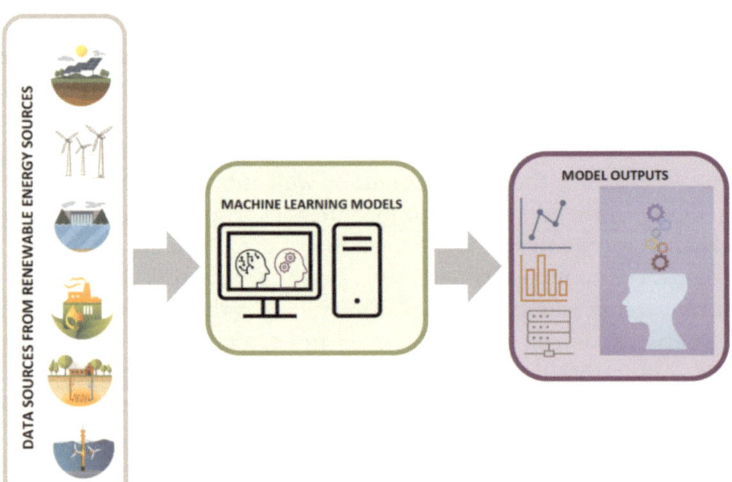


**Figure 3.**
*Structural layout of the article.*

information on the machine learning techniques and training algorithm types employed in studies related to renewable energy sources. Additionally, we will discuss the performance metrics used in ML models. Secondly, we will present a concise overview of each renewable energy source of interest. Lastly, we will list the existing literature on each renewable energy source of interest, arranged chronologically. Hence, we are certain that this article will assist readers engaged in literature studies and provide guidance for future research, owing to its comprehensive analysis of the role of machine learning methods in all forms of renewable energy.

## 2. Machine learning methods for renewable energy generation forecasting

### 2.1 Machine learning

ML refers to the capacity of a machine to make autonomous choices and provide outcomes to the user, achieved via the use of certain algorithms. It is often regarded as a sub-discipline of AI. Currently, ML is used for intricate data categorisation and the process of creating informed choices [16]. Simply, it involves creating algorithms that allow the system to acquire knowledge and make informed judgements. It is closely connected to mathematical optimisation, which offers techniques, principles, and practical applications to the subject. It is used in many computing jobs when it is not feasible to create and code explicit algorithms. ML is a discipline that involves developing algorithms based on input data and utilising output data to make predictions when new data becomes available. Data analytics encompasses the use of statistical and mathematical methodologies across several fields to create models, categorise, forecast, and cluster data. ML methods have been extensively used in several domains related to data-driven issues in recent years. ML approaches aim to identify correlations between input data and output data, regardless of whether they conform to mathematical problem structures [17]. The learning process is given as follows [18]:

$$R(w) = \int L[y, \phi(x, y, w)]p(x, y)dxdy \tag{1}$$

where $x$ and $y$ are the samples from the probability distribution $p$, $L$ is the loss function that balances the learning objectives, and the structure of the ML model is determined by $\phi(x, y, w)$ and $w$ parameters. After the ML models have been effectively trained using the training dataset, decision makers may get prediction output data by inputting the input data into the well-trained models [19]. ML may be categorised into three primary groups: supervised learning, unsupervised learning, and reinforcement learning (**Figure 4**) [20].

#### *2.1.1 Supervised learning*

Supervised learning methods are often used in the field of ML. Supervised learning is a kind of learning where a model is trained using a dataset that has been labelled. The model incorporates input and output parameters. Learning occurs based on the results provided for the corresponding input parameters. The algorithm repeatedly generates predictions on the training data. Learning concludes after the algorithm attains a satisfactory level of performance. In this kind of learning, the system is provided with a sample input and then associates it with the corresponding output.

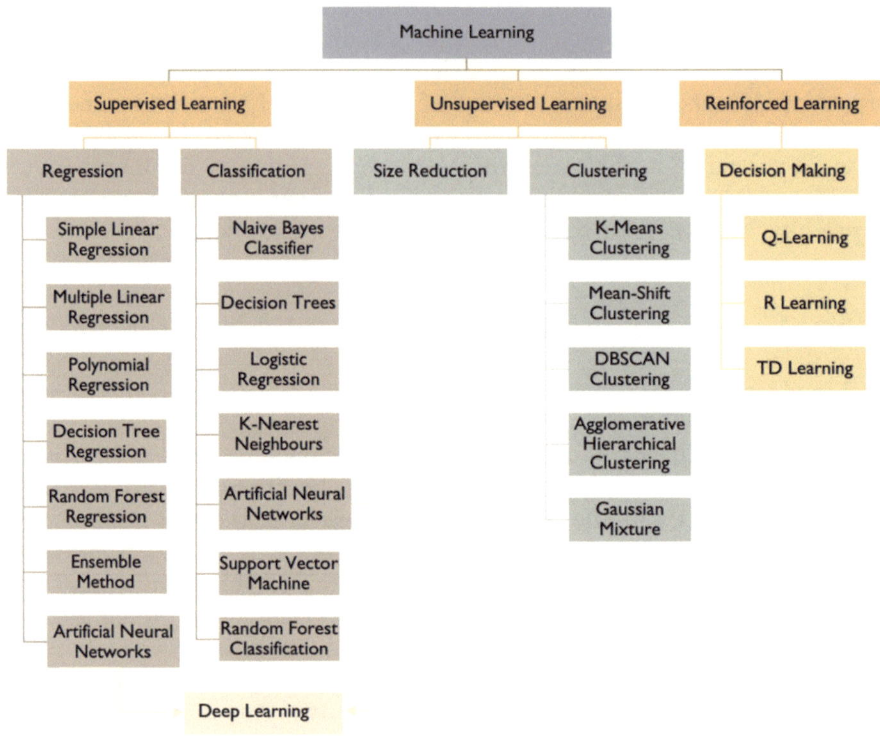


**Figure 4.**
*Machine learning categories and algorithms.*

In this learning paradigm, each sample is comprised of an input item, often represented as a vector, and a desired output value, which serves as a control signal. A supervised learning algorithm scrutinises and evaluates the training data to produce an inferential function that may be used to map fresh samples [21–24]. The ideal situation will enable the algorithm to accurately ascertain the class labels for occurrences that have not been previously seen. Generalising from the training data to scenarios that are not "reasonably" encountered by the learning algorithm is essential.

### *2.1.2 Unsupervised learning*

Unsupervised learning is another kind of machine learning in which an algorithm is trained on a dataset that does not have any known output variables. The goal is to discover patterns, structures, or connections within the data [25–27]. In unsupervised learning, the system is provided with sample inputs, but no output is produced. Classification is performed in order to accurately differentiate across data sets, since there is no specific targeted output in this case. Algorithms autonomously explore and reveal intriguing patterns within the data. Unsupervised learning methods extract a limited number of features from the data. Upon the introduction of new data, it leverages previously acquired characteristics to identify the data's class. Its primary use is for clustering and feature reduction. The challenge at hand is defining a function to uncover hidden patterns among unmarked data. As the samples or training sets provided to the learner lack labels, there is no mistake in the reward signal when assessing a potential solution.

*DOI: http://dx.doi.org/10.5772/intechopen.1007556*

#### *2.1.3 Reinforcement learning*

Reinforcement learning is a subfield of machine learning (ML) where an autonomous agent acquires the ability to make choices in a given environment with the goal of maximising a cumulative reward signal [28–30]. It focuses on determining the optimal actions for software agents to do in order to maximise the overall reward they get over time. It is extensively examined and used in several fields such as game theory, control theory, operations research, information theory, swarm intelligence, and static and genetic algorithms.

### 2.2 Artificial neural networks (ANN)

In 1943, McCulloch and Pitts proposed a mathematical model for a basic unit (neuron) of the human brain, which became the foundation for the theory of ANN [31]. One notable benefit of using ANN is their ability to create models from intricate natural systems with extensive inputs, resulting in improved usability and accuracy. The collective behaviour of an ANN is dictated by the interplay between the weights, inputs, and activation functions of each individual artificial neuron within the network. An ANN is composed of three layers: the input layer, the output layer, and the hidden layer. The input layer receives signals and data from the external environment. The hidden layer is an intermediary layer positioned between the input and output units, facilitating the interaction between these levels. The output layer is responsible for producing the final outcome of the system's operations. An artificial neuron consists of five parts: inputs, weights, addition function (combination function), activation function, and outputs. **Figure 5** depicts a feedforward ANN including $n$ input nodes representing a neuron, input variables $x_1, x_2, \dots, x_n$ weight values $w_1, w_2, \dots, w_n$ assigned to each input node, and the output of the neuron denoted as $o$. Input received by an artificial neuron from the external environment. This layer functions just to transmit received information without any modifications or additional responsibilities. Weights quantify the significance of incoming information on an artificial cell and its impact on the cell. Summation function determines the total input received by a cell. **Table 1** provides a list of some of the summation functions. The activation function is responsible for determining the output of a cell based on the input it receives, which is achieved by processing the net input to the cell. **Table 2** provides a list of several activation functions. The output value is governed by the activation function. This value is sent to the external environment or to another cell. The functioning of a cell in an ANN involves the following stages: (i) Inputs: An artificial neuron receives several inputs, with each input representing a distinct property or

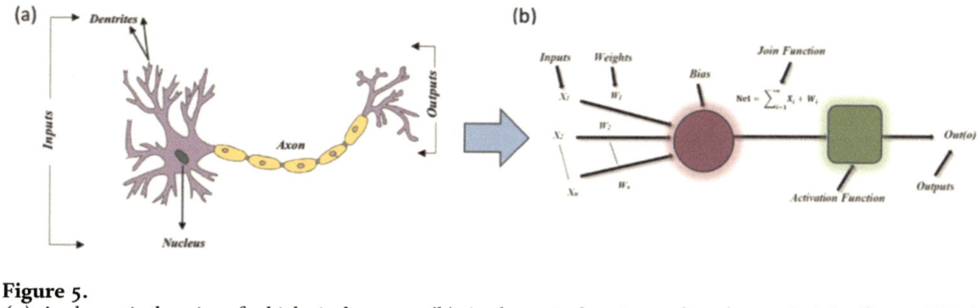


**Figure 5.**
*(a) A schematic drawing of a biological neuron. (b) A schematic drawing and working principle of an artificial neuron.*

| Function | Expression | Explanation |
|---|---|---|
| Total | $\text{Net} = \sum_{i=1}^{n} x_i * w_i$ | An elementary aggregation function that calculates the total of the outputs generated by each neuron inside a layer. This is often used in completely connected layers, where every neuron is linked to all neurons in the preceding layer. |
| Product | $\text{Net} = \prod_{i=1}^{n} x_i * w_i$ | The weight values are multiplied by the inputs, and the resulting values are then multiplied together to produce the net input. |
| Maximum | $\text{Net} = \max(x_i * w_i)$ | The net input of the artificial neuron is determined by multiplying the weights with the inputs among n inputs and selecting the greatest result. |
| Minimum | $\text{Net} = \min(x_i * w_i)$ | The net input of the artificial neuron is determined by multiplying the weights with the inputs from n inputs and selecting the least value. |
| Majority | $\text{Net} = \sum_{i=1}^{n} \text{sgn}(x_i * w_i)$ | The number of positive and negative weights is determined by multiplying the weights with the inputs from a total of n inputs. The bigger value is regarded as the net input of the cell. |
| Cumulative Total | $\text{Net} = (\text{old}) + \sum_{i=1}^{n} x_i * w_i$ | The cell primarily receives gathered information and calculates its net input by adding it to the previously received information in the cell. |

**Table 1.**
*The summation functions used in artificial neural networks [16, 17, 19].*

| Function | Expression | Explanation |
|---|---|---|
| Linear | $f(x) = x$ | The activation function is a basic function that directly maps its input to its output without introducing any nonlinearity. |
| Step | $f(x) = \begin{cases} 1 \text{ if } x >= 0, \\ 0 \text{ if } x < 0 \end{cases}$ | The activation function is a piecewise constant function that outputs a value of either 0 or 1 based on the sign of its input. |
| Sigmoid | $f(x) = \frac{1}{1+e^{-x}}$ | The activation function is a smooth and differentiable function that translates its input to a range between 0 and 1. |
| Hyperbolic Tangent | $f(x) = \tanh(x)$ | The activation function that maps any input value to the range [−1, 1]. |
| Threshold | $f(x) = \begin{cases} 1 \text{ if } x >= 0, \\ 0 \text{ if } x < 0 \end{cases}$ | The activation function is a piecewise constant function that produces an output of either 0 or 1 based on whether its input value is above or below the θ threshold. |
| Rectified Linear Unit (ReLU) | $f(x) = \max(0, x)$ | The (ReLU) activation function that returns 0 for negative input values and an input value for positive input values. |
| Leaky ReLU | $f(x) = \max(ax, x)$ | The ReLU variant with a small positive slope (α) for negative input values instead of 0. |
| Softmax | $f(x) = \frac{e^{x_i}}{\sum_{j=1}^{j} e^{x_j}}, \; i = 1, \ldots, j$ | The activation function that maps multiple input values to a probability distribution over several classes. |

**Table 2.**
*Activation functions used in artificial neural networks [16, 17, 19].*

part of the data being processed. (ii) Weighted sum: The inputs are multiplied by their respective weights, and the products are added together to produce a weighted total. The weights represent the significance of each input in relation to the output of the neuron. (iii) Activation function: The activation function processes the weighted sum and determines the output of the neuron. The sigmoid, hyperbolic tangent, and smoothed linear unit (ReLU) functions are typical instances of widely used activation functions. (iv) Result: The result of the artificial neuron serves as input for other neurons in the network, enabling the processing and transmission of information across the network. (v) Learning: The synaptic weights connecting neurons may be modified using a learning method like backpropagation. This enables the neural network to adapt and improve its predictions by minimising the discrepancy between the anticipated output and the actual output. To summarise, the operational mechanism of a cell in an ANN is computing a weighted total of the inputs, transmitting the total via an activation function to generate an output, and modifying the weights using a learning process.

## 2.3 Deep learning (DL)

DL refers to a subfield of machine learning that focuses on training ANNs to learn and make predictions or decisions. DL emulates the architecture and functionality of the human brain autonomously. It is a machine learning methodology used for the processing and analysis of algorithms and vast quantities of data. DL mostly relies on acquiring knowledge from the data's representation. The objective of DL is to construct models that have the ability to enhance data automatically without the need of explicit programming. Defining the first instance of DL in a specific person or group is challenging due to its gradual development via a collective endeavour including several academics such as Geoffrey Hinton, Yann LeCun, and Yoshua Bengio. DL originated in 1943 when Warren McCulloch and Walter Pitts used mathematical principles and algorithms to develop a computational system that mimics brain networks [32]. Hinton has provided a concise definition of DL and he describes it as a specific area within machine learning that use artificial neural networks with several layers to autonomously acquire knowledge and enhance performance via experience [33].

DL is the use of deep neural networks (DNN), which consist of numerous hidden layers, to analyse intricate patterns and connections within data. LeCun, a prominent figure in the field of DL and AI, provides a concise definition of deep learning. According to LeCun, deep learning is a specific area within machine learning that employs deep neural networks, which consist of numerous hidden layers, to effectively represent intricate patterns and connections within data [33]. LeCun states that deep learning enables computers to automatically extract advanced characteristics from raw data, allowing them to execute tasks like picture recognition and natural language processing with exceptional accuracy. The hierarchical representational learning in DL makes it highly suitable for complicated tasks like as image and voice recognition, machine translation, and other similar tasks. Every layer of the DNN is equipped with an activation function to generate a neuron response, while a biassed weighted sum provides the cost function. The mathematical representation for the DNN operation of two consecutive DNN layers [k − 1, k] is as follows:

$$y_i = f_i\left(\sum_{i=1}^{n} w_{ij} + b_j\right), i \in [0, n] \wedge i \in [0, m] \quad (2)$$

where $n$ and $m$ represent the number of layers in the $k-1$ and $k$th layers, respectively. When considering neuron $j$, the output is represented by $y_j$, while the input signal from neuron $i$ is denoted as $x_i$. Additionally, $b_j$ represents the bias of the neuron, and $w_{ij}$ represents the weight associated with the connections between neurons $i$ and $j$.

### *2.3.1 Convolutional neural network (CNN)*

Convolutional neural networks are mostly used for the purpose of image analysis. The concept was proposed using the animal vision system, as described by Hubel and Wiesel in 1968 [34]. It relies on the process of filtration. The filter will specify the properties that describe the characteristics of the picture. It yields favourable outcomes, particularly in classifier operations. Filters are used in various dimensions and magnitudes to expose low dominant characteristics [35, 36]. CNNs are a kind of artificial neural network that consists of several layers and are commonly employed in DL designs. LeCun et al. presented a gradient-based technique, which led to the creation of the first CNN network known as LeNet [37].

### *2.3.2 Recurrent neural network (RNN)*

The Simple Recurrent Network (SRN), first proposed by Elman, is a very popular neural network model for implicit learning [38]. The SRN is a neural network with three layers: input, hidden, and output. The system is taught to anticipate the subsequent stimulus by analysing the present input and the preceding inputs. The predictive capacity of the SRN is mostly derived from the additional set of context units that store a duplicate of the network's pattern of activity [38]. A recurrent neural network is a kind of ANN where the connections between units create a directed loop. This loop generates a network internal state that enables it to demonstrate dynamic temporal behaviour. In contrast to feedforward neural networks, RNNs have the ability to use their input memory to effectively handle arbitrary sequences of inputs [39]. The fundamental concept behind an RNN is to use sequential information. When dealing with image-based data, it is presumed that all inputs (or outputs) are mutually independent. However, this does not apply to disciplines that include temporal variables, such as natural language processing (NLP).

### *2.3.3 Long short-term memory (LSTM) networks*

When there are gaps in context between time strings in RNN systems, predicting the next string becomes very challenging [40], and RNNs are greatly disadvantaged in such cases. In 1997, Hochreiter and Schmidhuber proposed the use of LSTM to address this issue [41]. LSTM networks are indistinguishable from RNN networks. However, LSTM networks use a structure to compute the hidden state. LSTM is composed of memory cells. A cell is responsible for storing both the previous state and input information. These cells inside the network architecture determine the data to retain and the data to discard. During the subsequent stage, they merge the preceding state with the current memory and the incoming data. By using this method, it becomes feasible to preserve datasets by eliminating long-term dependencies.

### *2.3.4 Restricted Boltzmann machines*

Boltzmann machines are a kind of neural networks that were first developed in the late 1980s [42]. These techniques are rooted in statistical physics and use stochastic neurons, which sets them apart from most other neural network approaches. Restricted Boltzmann Machines (RBMs) are simpler variants of Boltzmann machines. A RBM is a kind of ANN that is capable of learning a probability distribution over its input data. It is a generative stochastic model. It gained popularity with the development of efficient learning algorithms by Geoffrey Hinton and his colleagues in the 2000s [43]. The connections between the hidden neurons at the top and the visible neurons at the bottom are eliminated in RBMs, which are a modification of the original Boltzmann machines. Only the synapses between the neurons in the visible layer and the hidden layer persist. Their weights are stored in the matrix $w$. RBMs are easier to understand compared to Boltzmann machines since they have fewer connections, making them more manageable for larger situations.

#### *2.3.5 Deep belief networks (DBN)*

The Deep Belief Network consists of a combination of many restricted Boltzmann models, each having hidden explanatory component neural networks that have numerous layers [44, 45]. The network layer serves as the intermediary between the levels, but it does not provide a direct link inside each layer. The training of the hidden layer captures the data dependence seen in the visible layer unit. This network design greatly reduces the intricacy of training and enables the practicality of deep learning. The advent of deep belief networks has been a significant milestone in the field of deep learning. These DBNs are now being widely used in different domains, including speech recognition and picture processing.

#### *2.3.6 Autoencoders (AE)*

An Autoencoder is primarily composed of three components: the encoder, the decoder, and the hidden layer [46]. An automated encoder first encodes the input signal and then uses the coded signal to rebuild the original signal. This encoded signal has the ability to reduce the discrepancy between the original signal and the reconstructed signal. During the encoding and reconstruction process, the encoder transforms the input data into a designated feature space. The decoder maps the features of the encoded signals back to the data space, resulting in the reconstruction of the original data. In the case of an automated encoder, the mapping often focuses on the input that will undergo encoding. If there is a discrepancy between the manipulated coding data and the input data, the system has the ability to regenerate the original signal in an altered format. Consequently, the characteristics are retrieved in order to facilitate automated learning.

### 2.4 Performance metrics for machine learning models

Various performance metrics have been verified in the literature, such as the correlation coefficient and root mean squared error, etc. When evaluating the effectiveness of an ML model, various metrics are commonly employed, and metrics help determine whether the model is performing well or not. When assessing the effectiveness and capabilities of ML models, researchers typically rely on a set of indicators. These indicators are calculated based on the summarised results, as shown in **Table 3**.

| Metric | Formula |
|---|---|
| Mean Error (ME) | $ME = \frac{\sum_{i=1}^{n} P_i}{n}$ |
| Mean Absolute Error (MAE) | $MAE = \frac{\sum_{i=1}^{n} \lvert P_i \rvert}{n}$ |
| Normalised Root Mean Square Error (NMAE) | $NMAE = \frac{1}{n}\sum_{i=1}^{n}\left\lvert\frac{P_i - A_i}{A_{\max}}\right\rvert$ |
| Mean Percentage Error (MPE) | $MPE = \frac{\sum_{i=1}^{n} P_i/A_i}{n/100}$ |
| Mean Absolute Percentage Error (MAPE) | $MAPE = \frac{100}{n}\sum_{i=1}^{n}\lvert P_i\rvert/\lvert A_i\rvert$ |
| Relative Mean Absolute Percentage Error (RMAPE) | $RMAPE = \frac{1}{n}\sum_{i=1}^{n}\left\lvert\frac{P_i - A_i}{A_i}\right\rvert \times 100$ |
| Maximum Absolute Relative Percentage Error (MARPE) | $MARPE = \max(P_i\%), i = 1, 2, \ldots, n$ |
| Mean Arctangent Absolute Percentage Error (MAAPE) | $MAAPE = \frac{1}{n}\sum_{i=1}^{n}\arctan\left(\left\lvert\frac{A_i - P_i}{A_i}\right\rvert\right)$ |
| Symmetric Mean Absolute Percentage Error (SMAPE) | $SMAPE = \frac{\sum_{i=1}^{n}\lvert P_i - A_i\rvert}{\sum_{i=1}^{n}(A_i + P_i)}$ |
| Mean Absolute Log-difference Error (MALE) | $MALE = 100 \times \frac{\sum_{i=1}^{n}\lvert \log P_i - \log A_i\rvert}{n}$ |
| Mean Absolute Bias Error (MABE) | $MABE = \frac{1}{n}\sum_{i=1}^{n}\lvert P_i - A_i\rvert$ |
| Mean Relative Error (MRE) | $MRE = \frac{1}{n}\sum_{i=1}^{n}\lvert A_i - P_i\rvert$ |
| Mean Squared Error (MSE) | $MSE = \frac{1}{n}\sum_{i=1}^{n}(P_i - A_i)^2$ |
| Mean Bias Error (MBE) | $MBE = \frac{1}{n}\sum_{i=1}^{n}(P_i - A_i)$ |
| Mean Absolute Scaled Error (MASE) | $MASE = \frac{\sum_{i=1}^{n}\lvert P_i - A_i\rvert}{\frac{n}{n-1}\sum_{i=2}^{n}\lvert A_i - A_{i-1}\rvert}$ |
| Relative Squared error (RSE) | $RSE = \frac{\sum_{i=1}^{n}(P_i - A_i)^2}{\sum_{i=1}^{n}(\bar{A} - A_i)^2}$ |
| Relative Absolute Error (RAE) | $RAE = \frac{\sum_{i=1}^{n}\lvert P_i - A_i\rvert}{\sum_{i=1}^{n}\lvert\bar{A} - A_i\rvert}$ |
| Root Mean Square Error (RMSE) | $RMSE = \sqrt{\frac{\sum_{i=1}^{n}(P_i - A_i)^2}{n}}$ |
| Normalised Root Mean Square Error (NRMSE) | $NRMSE = \frac{\sqrt{\frac{\sum_{i=1}^{n} P_i^2}{n}}}{\bar{A}}$ |
| Relative Root Mean Square Error (RRMSE) | $RRMSE = \frac{\sqrt{\frac{1}{n}\sum_{i=1}^{n}(P_i - A_i)^2}}{\frac{1}{n}\sum_{i=1}^{n} A_i} \times 100$ |
| Relative Root Mean Squared Percentage Error (RRMSPE) | $RRMSPE = \frac{1}{n}\sum_{i=1}^{n}\left\lvert\frac{P_i - A_i}{A_i}\right\rvert \times 100$ |
| Anomaly Correlation coefficient (ACC) | $ACC = \frac{\sum_{i=1}^{n}(P_i - \bar{P})(A_i - \bar{A})}{\sqrt{\sum_{i=1}^{n}(P_i - \bar{P})^2\sum_{i=1}^{n}(A_i - \bar{A})^2}}$ |

| Metric | Formula |
|---|---|
| Pearson Correlation Coefficient (PCC) | $PCC = \frac{\sum_{i=1}^{n}(P_i-\overline{P})(A_i-\overline{A})}{\sqrt{\sum_{i=1}^{n}(P_i-\overline{P})^2}\sqrt{\sum_{i=1}^{n}(A_i-\overline{A})^2}}$ |
| Average Percentage Error (APE) | $APE = \frac{1}{n}\sum_{i=1}^{n}\left(\frac{P_i-A_i}{A_i}\right) \times 100$ |
| Average Coverage Error (ACE) | $ACE = \frac{1}{n}\sum_{i=1}^{n}c_i \times 100\% - \text{PINC}$<br>$c_i$: indicator of prediction, PINC: prediction interval nominal confidence |
| Sum of Squared Error (SSE) | $SSE = \sum_{i=1}^{n}P_i^2$ |
| Correlation Coefficient (R) | $R = \frac{\sum_{i=1}^{n}(A_i-\overline{A}_i)(P_i-\overline{P}_i)}{\sqrt{\sum_{i=1}^{n}(A_i-\overline{A}_i)^2\sum_{i=1}^{n}(P_i-\overline{P}_i)^2}}$ |
| Coefficient of Determination ($R^2$) | $R^2 = 1 - \sum_{i=1}^{n}(P_i - A_i)^2/\sum_{i=1}^{n}(A_i - \overline{A})^2$ |

*Parameters* A, P, *and* n *represent the actual measurements, predictions, and number of data points, respectively [16, 17, 19].*

**Table 3.**
*Performance metrics for AI models.*

# 3. 2D TMOs as photocatalysts literature review on applying machine-learning techniques to renewable energy sources

## 3.1 Solar energy

Currently, renewable energy sources are used to produce electrical energy. In order to establish a sustainable and eco-friendly electrical system, it is essential to ensure that a significant proportion of energy output comes from renewable sources [47]. Solar energy has become a very promising renewable option for generating electricity. Solar energy, which consists of sunlight and solar heat radiation, may be captured using both synthetic and natural ways that are ecologically benign [48]. Sunlight may be categorised as a highly effective, cost-effective, and environmentally friendly energy source [49]. The composition of photovoltaic (PV) cells and the dynamic climatic circumstances, such as temperature, humidity, and cloudiness, have a significant influence on solar energy. The energy production is directly influenced by the fluctuations in these parameters. To ensure the creation of reliable and accurate models, it is important to explicitly define the functioning of the systems. Recently, mathematical linkages and occurrences in physical systems have been described using statistical and data-based nonlinear equations [50–54]. Hence, it is essential to use a robust renewable energy prediction model that can analyse historical meteorological data and optimise the data to facilitate informed decision-making in the energy industry. This model will also calculate power availability, aiding in grid stability and enabling optimum unit commitment and cost-effective distribution [55]. The prediction model's accuracy is contingent upon the characteristics used in its construction. To create a reliable solar energy forecasting model, it is necessary to establish the climatic variables inside the model. Projections for solar energy are made for both the near and far future. Short-term predictions are used for making decisions about the distribution of economic load, whereas medium and long-term forecasts are utilised for planning solar projections [56]. In order to develop an optimal solar forecasting model for different prediction timescales, it is necessary to use many forecasting models. AI techniques have several applications in solar energy, including control, management, difficulties in PV systems, irradiance projections, size, and site adaptation. We are of the opinion that these categories include the bulk of the issues faced in PV systems when ML has been used. In the literature, solar energy prediction models are categorised as either indirect or direct prediction models [57, 58]. Estimations of solar energy conducted at different time intervals may also be used to gauge the power output of the plant directly while considering the features of the panels. The methods used to anticipate solar energy may be categorised into three types: statistical [59, 60], physical [61], and artificial intelligence [62]. Artificial intelligence methods are often used in this field because to their ability to acquire knowledge, enhance their understanding, and fill in gaps in data [63]. **Table 4** represents prominent studies in the field of solar energy that incorporates machine learning methodologies.

| Ref. no. | ML Technique | Application | Forecasting Horizon | Parameters |
|---|---|---|---|---|
| [64] | CNN | Solar radiation prediction | hourly | MAE |
| [65] | MLP-NN | Solar power forecasting | daily | MAE, MRE, RMSE |
| [66] | ANN | PV power forecasting | hourly | MSE, MAE |
| [67] | ANN-LM | Solar radiation prediction | hourly | R, MSE |
| [68] | Fuzzy-SVM | Solar radiation prediction | daily | RMSE, MAE |
| [69] | SVR-RBF | Solar radiation prediction | daily, monthly | $R^2$, MAPE, MABE, RMSE, RRMSE |
| [70] | SVR-RBF | Solar radiation prediction | hourly | $R^2$, RMSE |
| [71] | BPNN | Solar irradiance prediction | daily | RMSE, $R^2$, MAE, MBE |
| [72] | BPNN | Solar radiation prediction | daily | RMSE |
| [73] | BPNN | Solar beam radiation prediction | monthly | RMSE, MBE |
| [74] | BPNN | Solar energy prediction | monthly | R, $R^2$ |
| [75] | BPNN | Global solar radiation prediction | daily | MPE, MBE, RMSE |
| [76] | SVM, RBFNN, AR | Solar power prediction | hourly | MAE, MAPE |
| [77] | SVR, BPNN | PV energy prediction | hourly | MBE, MAE, RMSE |
| [78] | SVM | Solar power prediction | daily | RMSE |
| [79] | SVM | Solar radiation prediction | hourly | RMSE, MAPE, R |
| [80] | ANN | Solar irradiance and solar PV parameters forecasting | hourly | R, $R^2$, RMSE, MAE |
| [81] | ANN | Solar energy production forecasting | daily | RMSE |
| [82] | DCNN | PV power forecasting | daily | MAE, MSE |
| [83] | SVM, ANN, DL, k-NN | Global solar radiation prediction | daily | MAPE, RMSE, $R^2$ |
| [84] | DL | Solar radiation prediction | monthly | MBE, MSE, RMSE, $R^2$ |
| [85] | NN, SVR, RF, MLR | PV power forecasting | hourly | MAE, MRE |
| [86] | SVR, RF | Solar power forecast | hourly | RMSE |
| [87] | CNN | Solar power forecast | hourly | MSE, RMSE |

| Ref. no. | ML Technique | Application | Forecasting Horizon | Parameters |
|---|---|---|---|---|
| [88] | DNN, RNN, LSTM | Solar thermal system prediction | hourly | RMSE, $R^2$ |
| [89] | LSTM, GRU | Solar power forecast | hourly, daily | MAE, RMSE, MAPE |
| [90] | ANN | PV module temperature prediction | daily | $R^2$, MAE, MSE, RMSE, MAPE |
| [91] | SVM, RBF | Solar power generation prediction | monthly | MSE |

*The nomenclature used for ML models is as follows; MLP: multilayer perceptron, NN: neural network, LM: Levenberg-Marquardt algorithm, SVM: support vector machines, SVR: support vector regression, RBF: radial basis function, BPNN: back propagation neural network, RBFNN: radial basis function neural network, DCNN: deep convolutional neural network, k-NN: k-nearest neighbour, MLR: multiple linear regression, RF: random forest, and GRU: gated recurrent unit.*

**Table 4.**
*List of studies for application of ML approaches in solar energy.*

## 3.2 Wind energy

The movement of the Earth and the uneven distribution of sunlight on its surface, with more sunlight falling on the equator than the poles, leads to the formation of wind. The use of wind as a prominent energy source, via the conversion of its kinetic energy into mechanical energy using windmills and wind turbines, has been happening for centuries and continues to this day. Wind energy is one of the most significant and potentially significant renewable energy sources [92–94]. The potential for wind energy to generate electricity exists throughout the entire day. As a result, it is particularly well-suited for systems that necessitate a consistent supply of energy. However, the utilisation of wind energy is not without its challenges. Investment costs are initially greater than those of competing power facilities. The second obstacle pertains to the transportation of wind turbines, which is a challenging task. Consequently, the location chosen for the establishment of wind power plants must exhibit considerable promise and undergo a thorough analysis of its energy potential. Furthermore, wind energy is most effective in elevated and remote locations, necessitating the use of transmission lines to link the produced electrical energy to the power grid. Consequently, these parameters ought to be considered in the process of identifying the optimal site for the installation of the wind power plant. Like solar energy, wind energy is susceptible to adverse meteorological conditions such as temperature, humidity, and solar radiation. Energy output is significantly impacted by the variability of these parameters. Research in the field of wind energy forecasting employs a multitude of analytical techniques. Research endeavours have employed statistical, data mining, and ML methodologies to analyse wind energy forecasts. The ability of machine learning algorithms to adapt to changes in the dataset is their primary purpose. Research pertaining to wind energy prediction can be categorised into two distinct domains: short-term and long-term wind energy prediction. Long-term wind power forecasting methods attempt to predict wind power over extended time periods, ranging from a few days to 1 year, whereas short-term wind power forecasting methods aim to predict wind power over shorter time periods, ranging from 1 hour to a few days. **Table 5** lists well-known research on machine learning approaches in wind energy.

| Ref. no. | ML Technique | Application | Forecasting Horizon | Parameters |
|---|---|---|---|---|
| [95] | SVR, ANN, RF, GBM | Wind forecasting | Hourly | NMAE, NRMSE |
| [96] | CNN | Wind speed forecasting | Daily | MAE, RMSE, MAPE, NMAE, MASE |
| [97] | RF, SVM, DT | Wind power forecasting | Daily | RMSE, MAPE |
| [98] | LR, SVR | Wind speed prediction | Minutely | MAE, RMSE |
| [99] | MLP, CNN, RNN | Wind speed prediction | Hourly | $R^2$ |
| [100] | EDCNN | Wind power forecasting | Hourly | MAE, NRMSE, MAPE |
| [101] | CNN | Wind speed forecasting | Daily | MAE, RMSE |
| [102] | RNN, LSTM | Wind power prediction | Hourly | MSE |
| [103] | ANN, SVR, GPR | Wind power prediction | Hourly | NRSE |
| [104] | CNN | Wind power prediction | Minutely | MSE, RMSE |
| [105] | LSTM | Wind power forecasting | Daily | MAE, RMSE, R |
| [106] | LSTM | Wind speed forecasting | Hourly | R, MSE, MAPE |
| [106] | LSTM | Wind power forecasting | Minutely | RMSE, MAE |
| [107] | SVM | Wind power forecasting | Hourly | $R^2$, RMSE, MAE, MAPE |
| [108] | LSSVM | Wind speed prediction | Minutely | MAE, MRE, RMSE, MAPE |
| [109] | BPNN | Wind power prediction | Yearly | RMSE |
| [110] | BPNN, RBFNN, ADALINE | Wind speed prediction | Hourly | RMSE |
| [111] | ANN, RBFNN | Wind power prediction | Hourly | NMAE, NRMSE |
| [112] | SVM, BPNN | Wind speed prediction | Daily | MSE |
| [113] | BPNN | Wind power prediction | Monthly | MSE |
| [114] | ARMA, NLN, ANN | Wind speed prediction | Hourly | RMSE |
| [115] | SVM | Wind speed forecasting | Hourly | MSE, MRE |
| [95] | SVR, ANN, GBM, RF | Wind forecasting | Hourly | RMSE, NMAE, NRMSE |
| [116] | RF | Wind power forecasting | Hourly | MAE, RMSE, NMAE, MAPE, MASE |
| [117] | DNN | Wind power forecasting | Hourly | MAE, RMSE |
| [118] | CNN | Wind forecasting | Monthly | SMAPE |
| [119] | CNN | Wind power forecasting | Hourly | ACE |
| [104] | DNN | Wind power prediction | Minutely | MSE, RMSE |
| [120] | CNN | Wind power forecasting | Hourly | MAE, NRMSE, MAPE |
| [121] | LSTM | Wind farm power output prediction | Hourly | MSE, RMSE |
| [122] | LSTM | Wind power forecasting | Hourly | NRMSE, NMAE |
| [123] | LSTM, ANN | Wind speed prediction | Minutely | RMSE |

| Ref. no. | ML Technique | Application | Forecasting Horizon | Parameters |
|---|---|---|---|---|
| [124] | CNN, LSTM | Wind power prediction | Monthly | RMSE |
| [125] | RNN, GRU, LSTM | Wind speed forecasting | Hourly | RMSE, MAPE, MAE |

*The nomenclature used for ML models is as follows; GBM: gradient boosting machine, DT: decision tree, EDCNN: enhanced deep convolutional neural network, GPR: Gaussian process regression, LSSVM: least square support vector machine, and NLN: neural logic network.*

**Table 5.**
*List of studies for application of ML approaches in wind energy.*

### 3.3 Hydroelectric energy

Hydroelectric energy is a technique for generating electricity through the natural flow of water, like a waterfall, or by regulating water movement with a man-made barrier on a river. Hydropower production is responsible for a significant 75% of the renewable sources in the global electricity mix [126]. Therefore, it is essential to improve the efficiency of hydropower production because of its vital role in addressing economic and environmental concerns, which are highly relevant in today's world. Operating hydropower systems with precision offers numerous benefits, such as maximising water resources, increasing renewable energy production, addressing the growing energy demand, minimising equipment losses, and prolonging equipment lifespan. However, improving hydropower efficiency is a difficult task. Having a deep understanding and closely monitoring all energy transformation processes in a hydroelectric plant is crucial for achieving a positive outcome. Predicting hydropower generation involves forecasting the amount of electricity that can be supplied to the electrical system. For accurate forecasting, it is essential to have precise input prediction based on hydrological and meteorological data. This approach considers the main source and net head, which is determined by subtracting hydraulic losses at the water intake, losses in the efficiency of the turbine-generator set, internal consumption of auxiliary services, and electrical losses to the grid, just like a specialist would. Improving the accuracy of the forecast allows for a more efficient optimisation process in determining the best design and parameters for maximising the use of water resources. Effectively harnessing the variables of the power generating process is the main challenge in achieving energy-efficient power production. When the generating units are operating at their peak efficiency, the optimised dispatch of generating units becomes vital and necessitates the use of a performance criteria. An analysis of the operation of a hydroelectric plant can be done in a watercourse, either individually or in a series, with the overall operation being classified as either run-of-river or storage reservoir. The energy generating process is divided into three stages: pre-operation, real-time, and post-operation. The pre-operation stages can be divided into long-term, medium-term, short-term, and real-time scheduling [8]. Using ML methods has proven to be a successful approach for tackling these challenges. Operating at peak efficiency often requires the use of complex models and multi-objective functions, especially when it comes to hydropower facilities with reservoirs that serve multiple purposes. Machine learning is highly effective in tackling these complex problems, as numerous methodologies have been developed to accurately and efficiently identify the best possible solution. In addition, the use of machine learning can assist in reducing the uncertainty related to prediction variables, which is a common obstacle that makes operational planning more complex. ML algorithms frequently

| Ref. no. | ML Technique | Application | Forecasting Horizon | Parameters |
|---|---|---|---|---|
| [127] | BDTR, DFR, BLR, NNR | Reservoir water level forecasting | Daily | $R^2$, MAE, RMSE, RAE, RSE |
| [128] | LSSVM | Water level prediction | Daily | RMSE, MAPE |
| [129] | SVM, RBNN, GRNN | Reservoir level prediction | Daily | MSE, MAE, R |
| [130] | BPNN, k-NN | Scheduling of hydropower plant | Hourly | RMSE |
| [131] | BPNN | Modelling of rainfall-runoff process | Daily | RMSE |
| [132] | ANN, MR | River flow prediction | Daily | RMSE |
| [133] | BPNN | Stream flow prediction | Monthly | SSE |
| [134] | LR | Hydropower generation projection | Yearly | $R^2$ |
| [135] | SVM | Flow forecasting | Daily | MSE, $R^2$ |
| [136] | SVM | Streamflow forecasting | Monthly | MAE, MRE, $R^2$ |
| [136] | SVM | Streamflow forecasting | Monthly | RMSE, MAE, MAPE |
| [137] | SVR | Hydropower consumption forecasting | Seasonal | MAPE, RMSE |
| [138] | SVR | Streamflow forecasting | Monthly | MAPE, RMSE |
| [139] | ANN | River streamflow forecasting | Daily | R, RMSE, MAE |
| [140] | ANN | Reservoir runoff forecasting | Daily | R, RMSE, MAPE |
| [141] | ANN | rainfall | Monthly | MAE, MSE |
| [142] | ANN | Reservoir storage | Monthly | RMSE, MSE, MAPE |
| [143] | ANN, LSTM | Hydropower reservoir level prediction | Hourly | R |
| [144] | CNN/RNN, SVR, RFR, ANN, ELM, WKNNR | Hydroelectric power generation forecasting | Yearly | R, RMSE, MSE, MAE |

*The nomenclature used for ML models is as follows; BDTR: boosting and decision tree regression, DFR: decision forest regression, BLR: binomial logistic regression, NNR: neural network regression, RBNN: radial basis neural network, GRNN: generalised regression neural network, MR: multiple regression, LR: linear regression, RFR: random forest regression, ELM: extreme learning machine, and WKNNR: weighted k-nearest neighbour regression.*

**Table 6.**
*List of studies for application of ML approaches in hydroelectric energy.*

offer more reliable predictions, thus mitigating this problem. **Table 6** presents significant research in the hydroelectric energy field that incorporates ML techniques.

## 3.4 Geothermal energy

Geothermal energy is the incredible power harnessed from deep within the Earth's core. The heat source is connected to the internal composition of our planet and the various physical phenomena taking place within. Although the Earth's crust contains abundant heat, it is unevenly distributed and often too deep to be harnessed for industrial use. The movement of heat from the Earth's interior to the surface, where it eventually dissipates, often goes unnoticed. It is evident that the presence of a geothermal gradient can be confirmed by observing the increase in rock temperature with depth. On average, this gradient amounts to 30°C per kilometre of depth. Some parts of the Earth's crust can be reached through drilling, and in these areas, the gradient is significantly higher than the average. This phenomenon happens when there are magma bodies located relatively close to the surface, typically within a few kilometres. These magma bodies are in the process of cooling and may still be in a fluid state or transitioning into a solid state, resulting in the release of heat. In other regions without magmatic activity, the accumulation of heat is a result of unique geological conditions in the crust, leading to abnormally high values in the geothermal gradient. To effectively harness and make use of such a significant amount of heat, a medium is needed to efficiently transport the heat to the depths below the Earth's surface. Typically, heat is transferred from deeper areas to shallower regions through conduction and then convection, with geothermal fluids serving as the medium in this scenario. These fluids are rainwater that has seeped into the Earth's crust, been heated by hot rocks, and collected in aquifers, sometimes at high pressures and temperatures (up to above 300°C). These aquifers, or reservoirs, play a crucial role in the majority of geothermal fields. ML approaches have been extensively utilised in geothermal applications, specifically in the realm of sensors and robots for the design, control, and optimisation of geothermal well drilling. By utilising computer simulations and advanced ML techniques, it becomes feasible to analyse the dynamics of geothermal reservoirs and gain insights into their influence on the advancement of geothermal energy. Numerical simulations are valuable tools for studying enhanced geothermal systems and the behaviour of geothermal reservoirs. The literature contains several studies on geothermal energy utilising ML, as shown in **Table 7**.

### 3.5 Biomass energy

Biomass energy is a clean and renewable source of power. In fact, biomass energy relies on hydrocarbon materials that can be derived from both animals and plants. There is a diverse range of biomass that can be transformed into bioenergy products which includes woody biomass, agricultural residues, aquatic biomass, animal

| Ref. no. | ML Technique | Application | Parameters |
|---|---|---|---|
| [145] | RF | Mapping geothermal potential | RMSE, NRMSE |
| [146] | LSTM | Geothermal energy prediction | Accuracy |
| [147] | ANN | Forecasting of penetration in geothermal systems | APE |
| [145] | RF | Mapping geothermal potential | RMSE, NRMSE |
| [148] | BPNN | Geothermal power prediction | PRMSE, $R^2$ |
| [149] | BPNN | Geothermal map generation | PRMSE, $R^2$ |
| [150] | BPNN | Pressure prediction in geothermal plant | RMSE, MPE |
| [151] | RF, GBR | Predicting drilling rate of geothermal wells | MAE, MAPE, RMSE, $R^2$ |
| [152] | DT, RF | Predicting heat output from geothermal system | RMSE, $R^2$ |

**Table 7.**
*A list of studies that examine the use of ML techniques in geothermal energy.*

biomass, industrial biomass, and mixtures. There are three different types of biomass energy; (i) gases such as methane and ethane; (ii) liquids like ethanol and biodiesel; and (iii) solids including biochar and activated carbon. These energies are recognised for their cost-effectiveness, renewable nature, and lower levels of emerging pollutants when compared to fossil fuels. Bioenergy is widely recognised as a sustainable alternative to fossil fuels. Nevertheless, the widespread use of biomass-based energy products is hindered by issues such as inconsistent feedstock, economic viability of conversion, and the reliability of the supply chain. ML techniques have been increasingly utilised in bioenergy systems to tackle the challenges. In the literature ML models have been extensively used in four main areas; predicting biomass properties, predicting the performance of biomass conversion processes, predicting biofuel properties and the performance of bioenergy end-use systems, and supply chain modelling and optimisation. ML models have been proven to be highly valuable in generating data that is difficult to measure directly. It has also shown promise in enhancing traditional models of biomass conversion and biofuel end-uses. Additionally, ML has the potential to overcome the limitations of traditional computing techniques when it comes to designing and optimising bioenergy supply chains. As listed in **Table 8**, the literature includes numerous studies that employ ML to harness biomass energy.

| Ref. no. | ML Technique | Application | Parameters |
|---|---|---|---|
| [153] | GBRT | Prediction of higher heating value of biomass materials | MAPE, RMSE, $R^2$ |
| [154] | LR, SVM, DTR, GBRT | Prediction of hydrogen production via biomass gasification | $R^2$, MSE, MAE |
| [155] | DTR, MLP | Prediction of the outcomes of biomass gasification | $R^2$, RMSE, NRMSE |
| [156] | PR | Prediction of pyro product | $R^2$ |
| [157] | DL | Modelling thermochemical conversion of biomass | $R^2$, RMSE, MSE, |
| [158] | ANN | Prediction of gas-liquid-solid product distribution after solid waste pyrolysis process | $R^2$, RMSE |
| [159] | RF | Prediction of carbon contents of biochar based on the pyrolysis data of lignocellulosic biomass | $R^2$, RMSE |
| [160] | ANN, SVM | Prediction models for product distribution and bio-oil heating value of biomass pyrolysis | MAE, $R^2$, RMSE |
| [161] | RF | Forest aboveground biomass estimation | $R^2$, RMSE |
| [162] | RF | Modelling above ground biomass of fast growing trees | $R^2$ |
| [163] | ET, RF | Predicting co-pyrolysis of coal and biomass | $R^2$, RMSE, MSE |
| [164] | ELM, SLFN | Prediction of higher heating value of biomass | $R^2$, RMSE |

*The nomenclature used for ML models is as follows; GBRT: gradient-boosted regression tree, DTR: decision tree regression, PR: polynomial regression, ET: extremely tree, and SLFN: single layer feedforward neural network.*

**Table 8.**
*A list of studies that examine the use of ML techniques in biomass energy.*

### 3.6 Marine-based energy

The study and exploration of the oceans have a long history, initially driven by commercial and military interests. It later evolved into a structured field that is part of a significant branch of Earth sciences. Throughout history, humans have taken advantage of and utilised the resources of the ocean while studying technologies that can be applied to marine environments. This has allowed us to gain a better understanding of the vast and diverse marine world through various sources and scales of spatial information. Research in marine science relies heavily on advanced research methods. Due to the rapid progress of technology and the increasing use of big data, the amount of Earth system data stored has surpassed tens of petabytes in recent years [165]. We need efficient and timely analysis and processing of a large amount of ocean observation and simulation data to create precise and up-to-date forecast data in a shorter timeframe. Emerging technologies, like ML, are being increasingly utilised in ocean research and development. These technologies have the potential to enhance and support traditional numerical forecasting models by addressing their limitations in certain areas of ocean forecasting. When it comes to certain prediction and forecasting problems, it can be challenging to accurately describe them using classical mathematical models and traditional ocean theories. This is particularly true for the regional ocean and climate mechanisms that we have not fully understood yet. Given the growing volume of data collected from the ocean, the application of data-driven ML technology has the potential to greatly benefit these fields. Various physical phenomena in the ocean have significant effects on the Earth's climate, marine

| Ref. no. | ML Technique | Application | Parameters |
|---|---|---|---|
| [166] | ANN | Forecasting of wave energy | $R^2$ |
| [167] | NN, AR | Forecasting of wave energy | GoF (goodness of fit) |
| [168] | ANN, SVM | Water-level prediction | RMSE, $R^2$ |
| [169] | SVR | Tidal current prediction | MAPE, RMSE, MARPE |
| [170] | RNN, LSTM | Forecasting ocean waves | MSE, RMSE, MAPE, MAAPE |
| [171] | SVR, ELM | Ocean wave height prediction | RMSE |
| [172] | ANN | Ocean wave energy forecasting | MALE, RMSE |
| [173] | ELM, CNN, RNN | Wave energy period forecasting | RMSE, MAE, RRMSPE, RMAPE |
| [174] | BPNN | Predicting sea level variations | RMSE |
| [175] | BPNN | Sea wave height prediction | MSE |
| [176] | BPNN | Wave parameters prediction | RMSE |
| [177] | GP, BPNN | Sea level prediction | $R^2$ |
| [178] | RNN, CNN | Sea surface salinity prediction | ACC, RMSE |
| [179] | ConvLSTM | Sea level anomaly forecasting | RMSE, PCC |
| [180] | CNN | Sea wave wind prediction | MAE, RMSE, MAPE |
| [181] | ANN | Prediction of daily sea surface temperature | RMSE, MAE |

*The nomenclature used for ML models is as follows; AR: autoregressive modelling, GP: Gaussian process, and ConvLSTM: convolutional long short-term memory.*

**Table 9.**
*A list of studies that examine the use of ML techniques in marine-based energy.*

ecology, and human activities. These include internal waves, heatwaves, and eddies. Conventional approaches rely on scientific expertise and utilise satellite data for identification purposes. The advancement of ML has introduced fresh perspectives for utilising data related to the marine environment. ML can be used in various areas related to marine-based energy sources, including maintenance, resource assessment, energy production optimisation, environmental monitoring, control systems, fault detection, supply chain management, energy storage, economic modelling, and climate impact analysis. ML techniques have been more extensively promoted in marine research and numerous studies in the literature are enumerated in **Table 9**.

## 4. Conclusions and future perspectives

Given the growing concerns surrounding climate change and global warming, there has been a significant surge in the demand for renewable energy. The world anticipates a rapid transition to renewable energy sources in the near future in order to reduce carbon emissions from global electricity generation. Therefore, precise forecasting of renewable energy generation is of utmost importance, and numerous studies have been carried out in this field. Furthermore, the intricate nature of different environmental factors in renewable energy generation systems has made it unsuitable to rely solely on mathematical equations to describe them. Thus, the use of machine-learning models has become increasingly common in making predictions for renewable energy. This study examined machine-learning models used for renewable energy predictions in recent years. The findings of this study can be summarised as follows. Recently, there has been a significant rise in the utilisation of machine-learning techniques in the field of renewable energy. It is worth noting that the majority of discussions surrounding machine-learning technology in renewable-energy predictions have primarily centred around solar or wind energy forecasting. Therefore, in addition to forecasting solar and wind energy, it may be worth exploring other forms of renewable energy, such as tidal energy, biomass energy, wave energy, hydraulic power, and geothermal energy, for future research. Thus, our study serves as a valuable guide for researchers by encompassing machine learning applications across various renewable energy sources and highlighting significant studies in the literature. Potential advancements in machine learning in the energy era encompass a wide range of possibilities, including but not limited to:

- Analysing data to predict performance through the use of knowledge.
- It is necessary to explore new knowledge through transfer learning in order to expand the boundaries of the database and enhance the resilience of machine learning.
- ML in model predictive control models involve a balance between the time it takes to make decisions and the ability to make decisions in real-time.
- Predicting future national energy structures in line with carbon neutrality goals, along with strategies for achieving sustainable development objectives.
- Machine learning is utilised to expedite the discovery and synthesis of advanced materials for the purpose of energy conversion and storage.
- Implementing automatic fault detection and diagnosis on integrated systems

Furthermore, the utilisation of artificial intelligence techniques and hybrid models shows great potential in predicting the future of renewable energy. Additionally, the methods used for data preprocessing have a significant impact on the accuracy of machine-learning models when making predictions in the field of renewable energy. Therefore, exploring data preprocessing techniques and machine-learning models for predicting renewable energy could be a promising avenue for future research. Parameter selection plays a crucial role in determining the performance of machine-learning models in predicting renewable energy. Therefore, to enhance the performance of machine-learning models in predicting renewable energy, researchers can explore new metaheuristics like algorithms for selecting machine-learning parameters [182]. Based on these facts given, our study outlines a significant direction for future research and paves the way for prospective advancements in the area of renewable energy, as well as the existing developments in machine learning in this subject.

## Acknowledgements

Ö. C. Yelgel and C. Yelgel acknowledge the support from the University of Manchester, the National Graphene Institute and the School of Physics & Astronomy.

## Conflict of interest

The authors declare no conflict of interest.

## Author details

Övgü Ceyda Yelgel[1,2]* and Celal Yelgel[2,3]

1 Department of Electrical-Electronics Engineering, Recep Tayyip Erdoğan University, Rize, Türkiye

2 The Computational Science and Machine Learning Laboratory, Recep Tayyip Erdoğan University, Rize, Türkiye

3 Department of Electricity and Energy, Recep Tayyip Erdoğan University, Rize, Türkiye

*Address all correspondence to: oceyda.yelgel@erdogan.edu.tr

## References

[1] Wai RJ, Wang WH, Lin CY. High-performance stand-alone photovoltaic generation system. IEEE Transactions on Industrial Electronics. 2008;**55**:240

[2] Chen S, Liu P, Li Z. Low carbon transition pathway of power sector with high penetration of renewable energy. Renewable and Sustainable Energy Reviews. 2020;**130**:109985

[3] Azad AK, Rasul MG, Islam R, Shishir IR. Analysis of wind energy prospect for power generation by three Weibull distribution methods. Energy Procedia. 2015;**75**:722

[4] Executive Summary – Renewables 2023 – Analysis - IEA. Available from: https://www.iea.org/reports/renewables-2023/executive-summary.

[5] Du P, Wang J, Guo Z, Yang W. Research and application of a novel hybrid forecasting system based on multi-objective optimization for wind speed forecasting. Energy Conversion and Management. 2017;**150**:90

[6] Liu H, Tian HQ, Li YF, Zhang L. Comparison of four Adaboost algorithm based artificial neural networks in wind speed predictions. Energy Conversion and Management. 2015;**92**:67

[7] Guermoui M, Melgani F, Gairaa K, Mekhalfi ML. A comprehensive review of hybrid models for solar radiation forecasting. Journal of Cleaner Production. 2020;**258**:120357

[8] Bordin C, Skjelbred HI, Kong J, Yang Z. Machine learning for hydropower scheduling: State of the art and future research directions. Procedia Computer Science. 2020;**176**:1659

[9] Tagle-Salazar PD, Nigam KDP, Rivera-Solorio CI. Parabolic trough solar collectors: A general overview of technology, industrial applications, energy market, modeling, and standards. Green Processing and Synthesis. 2020;**9**:595

[10] Zhang Y, Pan G, Chen B, Han J, Zhao Y, Zhang C. Short-term wind speed prediction model based on GA-ANN improved by VMD. Renewable Energy. 2020;**156**:1373

[11] Jiang P, Liu Z, Niu X, Zhang L. A combined forecasting system based on statistical method, artificial neural networks, and deep learning methods for short-term wind speed forecasting. Energy. 2021;**217**:119361

[12] Li W, Jia X, Li X, Wang Y, Lee J. A Markov model for short term wind speed prediction by integrating the wind acceleration information. Renewable Energy. 2021;**164**:242

[13] Wickramasinghe L, Ekanayake P, Jayasinghe J. Machine learning and statistical techniques for daily wind energy prediction. Gazi University Journal of Science. 2022;**35**:1359

[14] Ziane A, Necaibia A, Sahouane N, Dabou R, Mostefaoui M, Bouraiou A, et al. Photovoltaic output power performance assessment and forecasting: Impact of meteorological variables. Solar Energy. 2021;**220**:745

[15] Sambana B, Appala Naidu P, Jarabala R, Narayana VNSL. Performance validation of wind turbines using machine learning methodologies. Materials Today Proceedings. 2023;**80**:2845

[16] Mitchell T, Buchanan B, DeJong G, Dietterich T, Rosenbloom P, Waibel A. Machine learning. Annual Review of Computer Science. 1990;**4**:417

[17] Rudin C, Wagstaff KL. Machine learning for science and society. Machine Learning. 2014;**95**:1

[18] Brunton SL, Noack BR, Koumoutsakos P. Machine learning for fluid mechanics. Annual Review of Fluid Mechanics. 2020;**52**:477

[19] Russell SJ et al. Artificial intelligence: A modern approach. Pearson Education Limited. 2016;**3**:1151

[20] Qiu J,Wu Q, Ding G, Xu Y, Feng S. A survey of machine learning for big data processing. EURASIP Journal on Advances in Signal Processing. 2016;**133**:1

[21] Liu B. Web data mining: Exploring hyperlinks, contents, and usage data. Data-Centric Systems and Applications. 2011;**1**:63

[22] Miorelli R, Kulakovskyi A, Chapuis B, D'Almeida O, Mesnil O. Supervised learning strategy for classification and regression tasks applied to aeronautical structural health monitoring problems. Ultrasonics. 2021;**113**:106372

[23] Sen PC, Hajra M, Ghosh M. Supervised classification algorithms in machine learning: A survey and review. Advances in Intelligent Systems and Computing. 2020;**937**:99

[24] Knudsen EI. Supervised learning in the brain. The Journal of Neuroscience. 1994;**14**:3985

[25] Dike HU, Zhou Y, Deveerasetty KK, Wu Q. Unsupervised learning based on artificial neural network: A review. In: 2018 IEEE International Conference on Cyborg Bionic Systems CBS; Shenzhen, China. IEEE; 2018. p. 322

[26] Glielmo A, Husic BE, Rodriguez A, Clementi C, Noé F, Laio A. Unsupervised learning methods for molecular simulation data. Chemical Reviews. 2021;**121**:9722

[27] Karhunen J, Raiko T, Cho KH. Unsupervised deep learning: A short review. Advances in Independent Component Analysis and Learning Machines. 2015;**7**:125

[28] Arulkumaran K, Deisenroth MP, Brundage M, Bharath AA. Deep reinforcement learning: A brief survey. IEEE Signal Processing Magazine. 2017;**34**:26

[29] Botvinick M, Ritter S, Wang JX, Kurth-Nelson Z, Blundell C, Hassabis D. Reinforcement learning, fast and slow. Trends in Cognitive Sciences. 2019;**23**:408

[30] Buşoniu L, Ernst D, De Schutter B, Babuška R. Approximate reinforcement learning: An overview. In: IEEE SSCI 2011: Symposium Series on Computational Intelligence - ADPRL 2011: 2011 IEEE Symposium on Adaptive Dynamic Programming and Reinforcement Learning; Paris, France. IEEE; 2011. p. 1

[31] Park YS, Lek S. Artificial neural networks: Multilayer perceptron for ecological modeling. Developments in Environmental Modelling. 2016;**28**:123

[32] McCulloch WS, Pitts W. A logical calculus of the ideas immanent in nervous activity. The Bulletin of Mathematical Biophysics. 1943;**5**:115

[33] Lecun Y, Bengio Y, Hinton G. Deep learning. Nature. 2015;**521**:436

[34] Hubel DH, Wiesel TN. Receptive fields and functional architecture of monkey striate cortex. The Journal of Physiology. 1968;**195**:215

[35] Fukushima K. Neocognitron: A self-organizing neural network model for a

mechanism of pattern recognition unaffected by shift in position. Biological Cybernetics. 1980;**36**:193

[36] Simard PY, Platt JC, Steinkraus D. Best practices for convolutional neural networks applied to visual document analysis. 2003;**3**:1

[37] LeCun Y, Bottou L, Bengio Y, Haffner P. Gradient-based learning applied to document recognition. Proceedings of the IEEE. 1998;**86**:2278

[38] Elman JL. Finding structure in time. Cognitive Science. 1990;**14**:179

[39] Tomáš T, Mikolov T. Overview introduction model description ASR results extensions MT results comparison main outcomes future work recurrent neural network based language model. 2010;**2**:1045

[40] Bengio Y, Simard P, Frasconi P. Learning long-term dependencies with gradient descent is difficult. IEEE Transactions on Neural Networks. 1994;**5**:157

[41] Hochreiter S, Schmidhuber J. Long short-term memory. Neural Computation. 1997;**9**:1735

[42] Information Processing in Dynamical Systems: Foundations of Harmony Theory. Available from: https://apps.dtic.mil/sti/citations/tr/ADA620727

[43] Salakhutdinov R, Mnih A, Hinton G. Restricted Boltzmann machines for collaborative filtering. ACM International Conference Proceeding Series. 2007;**227**:791

[44] Rifai S, Bengio Y, Courville A, Vincent P, Mirza M. Disentangling factors of variation for facial expression recognition. Lecture Notes in Computer Science. 2012;**7577**:808

[45] Ranzato M, Susskind J, Mnih V, Hinton G. On deep generative models with applications to recognition. In: Proceedings of the IEEE Computer Society Conference on Computer Vision and Pattern Recognition; Colorado Springs, CO, USA. IEEE; 2011. p. 2857

[46] Bengio Y. Learning deep architectures for AI. Foundations and Trends in Machine Learning. 2009;**2**:1

[47] Sen S, Ganguly S. Opportunities, barriers and issues with renewable energy development – A discussion. Renewable and Sustainable Energy Reviews. 2017;**69**:1170

[48] Kalogirou SA. Solar Energy Engineering: Processes and Systems. London: Academic Press; 2009. ISBN-10: 0123745012

[49] López-Alvarez M, Flores-Tlacuahuac A, Ricardez-Sandoval L, Rivera-Solorio C. Optimal start-up policies for a solar thermal power plant. Industrial and Engineering Chemistry Research. 2018;**57**:1026

[50] Xing J, Luo K, Wang H, Gao Z, Fan J. A comprehensive study on estimating higher heating value of biomass from proximate and ultimate analysis with machine learning approaches. Energy. 2019;**188**:116077

[51] Ahmad MW, Reynolds J, Rezgui Y. Predictive modelling for solar thermal energy systems: A comparison of support vector regression, random forest, extra trees and regression trees. Journal of Cleaner Production. 2018;**203**:810

[52] Miozzo M, Zordan D, Dini P, Rossi M. SolarStat: Modeling photovoltaic

sources through stochastic Markov processes. In: ENERGYCON 2014 - IEEE International Energy Conference; Cavtat, Croatia. IEEE; 2014. p. 688

[53] Chow TT. Performance analysis of photovoltaic-thermal collector by explicit dynamic model. Solar Energy. 2003;**75**:143

[54] López-Bautista AO, Flores-Tlacuahuac A, Nigam KDP. Optimal start-up policies for a nanofluid-based solar thermal power plant. Industrial and Engineering Chemistry Research. 2019; **58**:19135

[55] De Giorgi MG, Congedo PM, Malvoni M. Photovoltaic power forecasting using statistical methods: Impact of weather data. IET Science, Measurement & Technology. 2014;**8**:90

[56] Wan C, Zhao J, Song Y, Xu Z, Lin J, Hu Z. Photovoltaic and solar power forecasting for smart grid energy management. CSEE Journal of Power and Energy Systems. 2016;**1**:38

[57] Das UK, Tey KS, Seyedmahmoudian M, Mekhilef S, Idris MYI, Van Deventer W, et al. Forecasting of photovoltaic power generation and model optimization: A review. Renewable and Sustainable Energy Reviews. 2018;**81**:912

[58] Antonanzas J, Osorio N, Escobar R, Urraca R, Martinez-de-Pison FJ, Antonanzas-Torres F. Review of photovoltaic power forecasting. Solar Energy. 2016;**136**:78

[59] Voyant C, Notton G, Kalogirou S, Nivet ML, Paoli C, Motte F, et al. Machine learning methods for solar radiation forecasting: A review. Renewable Energy. 2017;**105**:569

[60] Yadav HK, Pal Y, Tripathi MM. Photovoltaic power forecasting methods in smart power grid. In: 12th IEEE International Conference Electronics, Energy, Environment, Communication, Computer, Control: (E3-C3), INDICON; New Delhi, India. IEEE; 2015. p. 1

[61] Raza MQ, Nadarajah M, Ekanayake C. On recent advances in PV output power forecast. Solar Energy. 2016; **136**:125

[62] Monjoly S, André M, Calif R, Soubdhan T. Hourly forecasting of global solar radiation based on multiscale decomposition methods: A hybrid approach. Energy. 2017;**119**:288

[63] Benali L, Notton G, Fouilloy A, Voyant C, Dizene R. Solar radiation forecasting using artificial neural network and random forest methods: Application to normal beam, horizontal diffuse and global components. Renewable Energy. 2019;**132**:871

[64] Kuo PH, Huang CJ. A green energy application in energy management systems by an artificial intelligence-based solar radiation forecasting model. Energies. 2018;**11**:819

[65] Rafati A, Joorabian M, Mashhour E, Shaker HR. High dimensional very short-term solar power forecasting based on a data-driven heuristic method. Energy. 2021;**219**:119647

[66] Sun L, Sun Y. Photovoltaic power forecasting based on artificial neural network and ultraviolet index. International Journal of Computing. 2022;**21**:153

[67] Choudhary A, Pandey D, Bhardwaj S. Global solar radiation estimation modeling using artificial neural network: A case study on metro cities of India. Lecture Notes in Electrical Engineering. 2021;**702**:479

[68] Baser F, Demirhan H. A fuzzy regression with support vector machine approach to the estimation of horizontal global solar radiation. Energy. 2017;**123**:229

[69] Mohammadi K, Shamshirband S, Anisi MH, Amjad Alam K, Petković D. Support vector regression based prediction of global solar radiation on a horizontal surface. Energy Conversion and Management. 2015;**91**:433

[70] Piri J, Shamshirband S, Petković D, Tong CW, Rehman MHU. Prediction of the solar radiation on the earth using support vector regression technique. Infrared Physics & Technology. 2015;**68**:179

[71] Mellit A, Pavan AM. A 24-h forecast of solar irradiance using artificial neural network: Application for performance prediction of a grid-connected PV plant at Trieste, Italy. Solar Energy. 2010;**84**:807

[72] Rehman S, Mohandes M. Artificial neural network estimation of global solar radiation using air temperature and relative humidity. Energy Policy. 2008;**36**:571

[73] Alam S, Kaushik SC, Garg SN. Computation of beam solar radiation at normal incidence using artificial neural network. Renewable Energy. 2006;**31**:1483

[74] Fadare DA. Modelling of solar energy potential in Nigeria using an artificial neural network model. Applied Energy. 2009;**86**:1410

[75] Jiang Y. Computation of monthly mean daily global solar radiation in China using artificial neural networks and comparison with other empirical models. Energy. 2009; **34**:1276

[76] Zeng J, Qiao W. Short-term solar power prediction using a support vector machine. Renewable Energy. 2013;**52**:118

[77] Li Z, Mahbobur Rahman SM, Vega R, Dong B. A hierarchical approach using machine learning methods in solar photovoltaic energy production forecasting. Energies. 2016;**9**:55

[78] Sharma N, Sharma P, Irwin D, Shenoy P. Predicting solar generation from weather forecasts using machine learning. In: 2011 IEEE International Conference on Smart Grid Communications, SmartGridComm. Brussels, Belgium: IEEE; 2011. p. 528

[79] Belaid S, Mellit A, Boualit H, Zaiani M. Hourly global solar forecasting models based on a supervised machine learning algorithm and time series principle. International Journal of Ambient Energy. 2022;**43**:1707

[80] Bamisile O, Oluwasanmi A, Obiora S, Osei-Mensah E, Asoronye G, Huang Q. Application of deep learning for solar irradiance and solar photovoltaic multi-parameter forecast. Energy Sources, Part A: Recovery, Utilization, and Environmental Effects. 2020:1-21. DOI: 10.1080/15567036.2020.1801903

[81] Juan S, Parra Raffán LC, Romero A, Martinez M. Solar energy production forecasting through artificial neuronal networks, considering the Föhn, North and South winds in San Juan, Argentina. The Journal of Engineering. 2019;**18**:4824

[82] Zang H, Cheng L, Ding T, Cheung KW, Wei Z, Sun G. Day-ahead photovoltaic power forecasting approach based on deep convolutional neural networks and meta learning. International Journal of Electrical Power & Energy Systems. 2020;**118**: 105790

[83] Ağbulut Ü, Gürel AE, Biçen Y. Prediction of daily global solar radiation using different machine learning algorithms: Evaluation and comparison. Renewable and Sustainable Energy Reviews. 2021;**135**:110114

[84] Üstün İ, Üneş F, Mert İ, Karakuş C. A comparative study of estimating solar radiation using machine learning approaches: DL, SMGRT, and ANFIS. Energy Sources, Part A: Recovery, Utilization, and Environmental Effects. 2022;**44**:10322

[85] Rana M, Rahman A. Multiple steps ahead solar photovoltaic power forecasting based on univariate machine learning models and data re-sampling. Sustainable Energy, Grids and Networks. 2020;**21**:100286

[86] Bajpai A, Duchon M. A hybrid approach of solar power forecasting using machine learning. In: Proceedings - 2019 3rd International Conference on Smart Grid and Smart Cities, ICSGSC 2019. Berkeley, USA: IEEE; 2019. p. 108

[87] Sun Y, Venugopal V, Brandt AR. Short-term solar power forecast with deep learning: Exploring optimal input and output configuration. Solar Energy. 2019;**188**:730

[88] Correa-Jullian C, Cardemil JM, López Droguett E, Behzad M. Assessment of deep learning techniques for prognosis of solar thermal systems. Renewable Energy. 2020;**145**:2178

[89] Wen L, Zhou K, Yang S, Lu X. Optimal load dispatch of community microgrid with deep learning based solar power and load forecasting. Energy. 2019;**171**:1053

[90] Keddouda A, Ihaddadene R, Boukhari A, Atia A, Arıcı M, Lebbihiat N, et al. Photovoltaic module temperature prediction using various machine learning algorithms: Performance evaluation. Applied Energy. 2024;**363**:123064

[91] Aldosari O, Batiyah S, Elbashir M, Alhosaini W, Nallaiyagounder K. Performance evaluation of multiple machine learning models in predicting power generation for a grid-connected 300 MW solar farm. Energies. 2024;**17**:525

[92] Jung J, Broadwater RP. Current status and future advances for wind speed and power forecasting. Renewable and Sustainable Energy Reviews. 2014; **31**:762

[93] Wang X, Guo P, Huang X. A review of wind power forecasting models. Energy Procedia. 2011;**12**:770

[94] Chang W-Y, Chang W-Y. A literature review of wind forecasting methods. Journal of Power and Energy Engineering. 2014;**2**:161

[95] Feng C, Cui M, Hodge BM, Zhang J. A data-driven multi-model methodology with deep feature selection for short-term wind forecasting. Applied Energy. 2017;**190**:1245

[96] Hong YY, Satriani TRA. Day-ahead spatiotemporal wind speed forecasting using robust design-based deep learning neural network. Energy. 2020;**209**: 118441

[97] Chaudhary A, Sharma A, Kumar A, Dikshit K, Kumar N. Short term wind power forecasting using machine learning techniques. Journal of Statistics and Management Systems. 2020;**23**:145

[98] Ali MEK, Hassan MZ, Ali ABMS, Kumar J. Prediction of wind speed using real data: An analysis of statistical machine learning techniques. In:

Proceedings - 2017 4th Asia-Pacific World Congress on Computer Science and Engineering, APWC on CSE 2017. Mana Island, Fiji: IEEE; 2017. p. 259

[99] Manero J, Béjar J, Cortés U. Dust in the wind... Deep learning application to wind energy time series forecasting. Energies. 2019;**12**:2385

[100] Mujeeb S, Javaid N, Gul H, Daood N, Shabbir S, Arif A. Wind power forecasting based on efficient deep convolution neural networks. Lecture Notes in Networks and Systems. 2020; **96**:47

[101] Huang CJ, Kuo PH. A short-term wind speed forecasting model by using artificial neural networks with stochastic optimization for renewable energy systems. Energies. 2018;**11**:2777

[102] Dong D, Sheng Z, Yang T. Wind power prediction based on recurrent neural network with long short-term memory units. In: 2018 IEEE International Conference on Renewable Energy and Power Engineering, REPE 2018. Toronto, ON, Canada: IEEE; 2018. p. 34

[103] Sharifzadeh M, Sikinioti-Lock A, Shah N. Machine-learning methods for integrated renewable power generation: A comparative study of artificial neural networks, support vector regression, and Gaussian process regression. Renewable and Sustainable Energy Reviews. 2019; **108**:513

[104] Yu R, Liu Z, Li X, Lu W, Ma D, Yu M, et al. Scene learning: Deep convolutional networks for wind power prediction by embedding turbines into grid space. Applied Energy. 2019;**238**:249

[105] Yin R, Li D, Wang Y, Chen W. Forecasting method of monthly wind power generation based on climate model and long short-term memory neural network. Global Energy Interconnection. 2020;**3**:571

[106] López G, Arboleya P. Short-term wind speed forecasting over complex terrain using linear regression models and multivariable LSTM and NARX networks in the Andes Mountains, Ecuador. Renewable Energy. 2022;**183**:351

[107] Li LL, Zhao X, Tseng ML, Tan RR. Short-term wind power forecasting based on support vector machine with improved dragonfly algorithm. Journal of Cleaner Production. 2020;**242**:118447

[108] Fu T, Wang C. A hybrid wind speed forecasting method and wind energy resource analysis based on a swarm intelligence optimization algorithm and an artificial intelligence model. Sustainability. 2018;**10**:3913

[109] Carolin Mabel M, Fernandez E. Analysis of wind power generation and prediction using ANN: A case study. Renewable Energy. 2008;**33**:986

[110] Li G, Shi J. On comparing three artificial neural networks for wind speed forecasting. Applied Energy. 2010;**87**:2313

[111] Sideratos G, Hatziargyriou ND. An advanced statistical method for wind power forecasting. IEEE Transactions on Power Systems. 2007;**22**:258

[112] Mohandes MA, Halawani TO, Rehman S, Hussain AA. Support vector machines for wind speed prediction. Renewable Energy. 2004;**29**:939

[113] Mabel MC, Fernandez E. Estimation of energy yield from wind farms using artificial neural networks. IEEE Transactions on Energy Conversion. 2009;**24**:459

[114] Sfetsos A. A comparison of various forecasting techniques applied to mean hourly wind speed time series. Renewable Energy. 2000;**21**:23

[115] Li Y, Yang P, Wang H. Short-term wind speed forecasting based on improved ant colony algorithm for LSSVM. Cluster Computing. 2019; **22**:11575

[116] Lahouar A, Ben Hadj Slama J. Hour-ahead wind power forecast based on random forests. Renewable Energy. 2017;**109**:529

[117] Khan M, Liu T, Ullah F. A new hybrid approach to forecast wind power for large scale wind turbine data using deep learning with TensorFlow framework and principal component analysis. Energies. 2019;**12**:2229

[118] Harbola S, Coors V. One dimensional convolutional neural network architectures for wind prediction. Energy Conversion and Management. 2019;**195**:70

[119] Wang HZ, Li GQ, Wang GB, Peng JC, Jiang H, Liu YT. Deep learning based ensemble approach for probabilistic wind power forecasting. Applied Energy. 2017;**188**:56

[120] Mujeeb S, Alghamdi TA, Ullah S, Fatima A, Javaid N, Saba T. Exploiting deep learning for wind power forecasting based on big data analytics. Applied Sciences. 2019;**9**:4417

[121] Eze EC, Chatwin CR. Enhanced recurrent neural network for short-term wind farm power output prediction. IJRDO-Journal of Applied Science. 2019; **5**:28

[122] Cali U, Sharma V. Short-term wind power forecasting using long-short term memory based recurrent neural network model and variable selection. International Journal of Smart Grid and Clean Energy. 2019;**8**:103

[123] Cardona JL, Howland MF, Dabiri JO. Seeing the wind: Visual wind speed prediction with a coupled convolutional and recurrent neural network. Advances in Neural Information Processing Systems. 2019;**33**:1

[124] Hanifi S, Cammarono A, Zare-Behtash H. Advanced hyperparameter optimization of deep learning models for wind power prediction. Renewable Energy. 2024;**221**:119700

[125] Ammar E, Xydis G. Wind speed forecasting using deep learning and preprocessing techniques. International Journal of Green Energy. 2024;**21**:988

[126] Turgeon K, Trottier G, Turpin C, Bulle C, Margni M. Empirical characterization factors to be used in LCA and assessing the effects of hydropower on fish richness. Ecological Indicators. 2021;**121**:107047

[127] Sapitang M, Ridwan WM, Kushiar KF, Ahmed AN, El-Shafie A. Machine learning application in reservoir water level forecasting for sustainable hydropower generation strategy. Sustainability. 2020;**12**:6121

[128] Guo T, He W, Jiang Z, Chu X, Malekian R, Li Z. An improved LSSVM model for intelligent prediction of the daily water level. Energies. 2019;**12**:112

[129] Üneş F, Demirci M, Taşar B, Kaya YZ, Varçin H. Estimating dam reservoir level fluctuations data-driven techniques. Polish Journal of Environmental Studies. 2019;**28**:3451

[130] Liang RH, Hsu YY. Scheduling of hydroelectric generations using artificial neural networks. IEEE Proceedings:

Generation, Transmission and Distribution. 1994;**141**:452

[131] Smith J, Eli RN. Neural-network models of rainfall-runoff process. Journal of Water Resources Planning and Management. 1995;**121**:499

[132] Firat M, Güngör M. River flow estimation using adaptive neuro fuzzy inference system. Mathematics and Computers in Simulation. 2007;**75**:87

[133] Dolling OR, Varas EA. Artificial neural networks for streamflow prediction. Journal of Hydraulic Research. 2002;**40**:547

[134] Kao SC, Sale MJ, Ashfaq M, Uria Martinez R, Kaiser DP, Wei Y, et al. Projecting changes in annual hydropower generation using regional runoff data: An assessment of the United States federal hydropower plants. Energy. 2015;**80**:239

[135] Jianzhong W, Ling L, Juan C. Combination of genetic algorithm and support vector machine for daily flow forecasting. In: Proceedings - 4th International Conference on Natural Computation, ICNC 2008. Vol. 2. JǏnan, China: IEEE; 2008. p. 31

[136] Guo J, Zhou J, Qin H, Zou Q, Li Q. Monthly streamflow forecasting based on improved support vector machine model. Expert Systems with Applications. 2011;**38**:13073

[137] Wang S, Yu L, Tang L, Wang S. A novel seasonal decomposition based least squares support vector regression ensemble learning approach for hydropower consumption forecasting in China. Energy. 2011;**36**:6542

[138] Luo X, Yuan X, Zhu S, Xu Z, Meng L, Peng J. A hybrid support vector regression framework for streamflow forecast. Journal of Hydrology (Amst). 2019;**568**:184

[139] Tayebiyan A, Mohammad T, Ghazali AH, Mashohor S. Artificial neural network for modelling rainfall-runoff. Pertanika Journal of Science & Technology. 2016;**24**:319

[140] Cheng CT, Niu WJ, Feng ZK, Shen JJ, Chau KW. Daily reservoir runoff forecasting method using artificial neural network based on quantum-behaved particle swarm optimization. Water. 2015;**7**:4232

[141] Basnayake ND, Attygalle D, Liyanage LH, Nandalal KDW, Dilini WMN, Hansen LL. Ensemble forecast for monthly reservoir inflow; a dynamic neural network approach. In: Proceedings of the 4th Annual International Conference on Operations. Singapore; 2016. p. 18

[142] Abdulkadir T, Salami A, Sule BF, Adeyemo JA. Neural network based model for forecasting reservoir storage for hydropower dam operation. International Journal of Engineering Research and General Science. 2015;**3**:639

[143] Sinh Nguyen P, Huy Nguyen T, Hung Nguyen T, Ba JSCS, Thanh N, Chau H, et al. A real-time flood forecasting hybrid machine learning hydrological model for Krong H'nang hydropower reservoir. River. 2024;**3**:107

[144] Sahin ME, Ozbay KM. Smart hydropower management: Utilizing machine learning and deep learning methods to enhance dam's energy generation efficiency. Neural Computing and Applications. 2024;**36**:11195-11211

[145] Assouline D, Mohajeri N, Gudmundsson A, Scartezzini JL. A machine learning approach for mapping

the very shallow theoretical geothermal potential. Geothermal Energy. 2019;**7**:1

[146] Gangwani P, Soni J, Upadhyay H, Joshi S. A deep learning approach for modeling of geothermal energy prediction. International Journal of Computer Science and Information Security (IJCSIS). 2020;**18**:62

[147] Diaz MB, Kim KY, Kang TH, Shin HS. Drilling data from an enhanced geothermal project and its pre-processing for ROP forecasting improvement. Geothermics. 2018;**72**:348

[148] Arslan O. Power generation from medium temperature geothermal resources: ANN-based optimization of Kalina cycle system-34. Energy. 2011;**36**:2528

[149] Arslan O, Yetik O. ANN based optimization of supercritical ORC-binary geothermal power plant: Simav case study. Applied Thermal Engineering. 2011;**31**:3922

[150] Bassam A, Del Castillo ÁA, García-Valladares O, Santoyo E. Determination of pressure drops in flowing geothermal wells by using artificial neural networks and wellbore simulation tools. Applied Thermal Engineering. 2015;**75**:1217

[151] Yehia T, Gasser M, Ebaid H, Meehan N, Okoroafor ER. Comparative analysis of machine learning techniques for predicting drilling rate of penetration (ROP) in geothermal wells: A case study of FORGE site. Geothermics. 2024;**121**: 103028

[152] Nath F, Garcia Romero NA, Cabezudo E, Sutradhar S, Cisneros E, Portillo O, et al. Predicting future heat outputs from enhanced geothermal system utilizing machine learning approach. In: SPE Western Regional Meeting Proceedings. Palo Alto, California, USA; 2024. p. D021S010R005

[153] Samadi SH, Ghobadian B, Nosrati M. Prediction of higher heating value of biomass materials based on proximate analysis using gradient boosted regression trees method. Energy Sources, Part A: Recovery, Utilization, and Environmental Effects. 2021;**43**:672

[154] Ozbas EE, Aksu D, Ongen A, Aydin MA, Ozcan HK. Hydrogen production via biomass gasification, and modeling by supervised machine learning algorithms. International Journal of Hydrogen Energy. 2019;**44**:17260

[155] Elmaz F, Yücel Ö, Mutlu AY. Predictive modeling of biomass gasification with machine learning-based regression methods. Energy. 2020;**191**: 116541

[156] Potnuri R, Suriapparao DV, Sankar Rao C, Sridevi V, Kumar A. Effect of dry torrefaction pretreatment of the microwave-assisted catalytic pyrolysis of biomass using the machine learning approach. Renewable Energy. 2022;**197**:798

[157] Khan M, Raza Naqvi S, Ullah Z, Taqvi SAA, Nouman Aslam Khan M, Farooq W, et al. Applications of machine learning in thermochemical conversion of biomass—A review. Fuel. 2023;**332**: 126055

[158] Gu C, Wang X, Song Q, Li H, Qiao Y. Prediction of gas-liquid-solid product distribution after solid waste pyrolysis process based on artificial neural network model. International Journal of Energy Research. 2021;**45**:13786

[159] Zhu X, Li Y, Wang X. Machine learning prediction of biochar yield and carbon contents in biochar based on biomass characteristics and pyrolysis

conditions. Bioresource Technology. 2019;**288**:121527

[160] Chen X, Zhang H, Song Y, Xiao R. Prediction of product distribution and bio-oil heating value of biomass fast pyrolysis. Chemical Engineering and Processing - Process Intensification. 2018;**130**:36

[161] Dang ATN, Nandy S, Srinet R, Luong NV, Ghosh S, Senthil KA. Forest aboveground biomass estimation using machine learning regression algorithm in yok don National Park, Vietnam. Ecological Informatics. 2019;**50**:24

[162] Wongchai W, Onsree T, Sukkam N, Promwungkwa A, Tippayawong N. Machine learning models for estimating above ground biomass of fast growing trees. Expert Systems with Applications. 2022;**199**:117186

[163] Wei H, Luo K, Xing J, Fan J. Predicting co-pyrolysis of coal and biomass using machine learning approaches. Fuel. 2022;**310**:122248

[164] Dai Z, Chen Z, Selmi A, Jermsittiparsert K, Denić NM, Nešić Z. Machine learning prediction of higher heating value of biomass. Biomass Conversion and Biorefinery. 2023;**13**: 3659

[165] Agapiou A. Remote sensing heritage in a petabyte-scale: Satellite data and heritage earth engine applications. International Journal of Digital Earth. 2017;**10**:85

[166] Avila D, Marichal GN, Padrón I, Quiza R, Hernández Á. Forecasting of wave energy in Canary Islands based on artificial intelligence. Applied Ocean Research. 2020;**101**:102189

[167] Shi S, Patton RJ, Liu Y. Short-term wave forecasting using Gaussian process for optimal control of wave energy converters. IFAC-PapersOnLine. 2018; **51**:44

[168] Zhang S, Lu L, Yu J, Zhou H. Short-term water level prediction using different artificial intelligent models. In: 2016 5th International Conference on Agro-Geoinformatics. Tianjin, China: IEEE; 2016. p. 1

[169] Kavousi-Fard A, Su W. A combined prognostic model based on machine learning for tidal current prediction. IEEE Transactions on Geoscience and Remote Sensing. 2017;**55**:3108

[170] Pirhooshyaran M, Snyder LV. Forecasting, hindcasting and feature selection of ocean waves via recurrent and sequence-to-sequence networks. Ocean Engineering. 2020;**207**:107424

[171] Kumar NK, Savitha R, Al MA. Regional ocean wave height prediction using sequential learning neural networks. Ocean Engineering. 2017; **129**:605

[172] Bento PMR, Pombo JAN, Mendes RPG, Calado MRA, Mariano SJPS. Ocean wave energy forecasting using optimised deep learning neural networks. Ocean Engineering. 2021;**219**:108372

[173] Ali M, Prasad R, Xiang Y, Sankaran A, Deo RC, Xiao F, et al. Advanced extreme learning machines vs. deep learning models for peak wave energy period forecasting: A case study in Queensland, Australia. Renew. Energy. 2021;**177**:1031

[174] Makarynskyy O, Makarynska D, Kuhn M, Featherstone WE. Predicting sea level variations with artificial neural networks at Hillarys boat harbour, Western Australia. Estuarine, Coastal and Shelf Science. 2004;**61**:351

[175] Londhe SN, Panchang V. One-day wave forecasts based on artificial neural networks. Journal of Atmospheric and Oceanic Technology. 2006;**23**:1593

[176] Makarynskyy O, Makarynska D. Wave prediction and data supplementation with artificial neural networks. Journal of Coastal Research. 2007;**23**:951

[177] Ghorbani MAG, Makarynskyy O, Shiri J, Makarynska D. Genetic programming for sea level predictions in an island environment. The International Journal of Ocean and Climate Systems. 2010;**1**:27

[178] Song T, Wang Z, Xie P, Han N, Jiang J, Xu D. A novel dual path gated recurrent unit model for sea surface salinity prediction. Journal of Atmospheric and Oceanic Technology. 2020;**37**:317

[179] Wang G, Wang X, Wu X, Liu K, Qi Y, Sun C, et al. A hybrid multivariate deep learning network for multistep ahead sea level anomaly forecasting. Journal of Atmospheric and Oceanic Technology. 2022;**39**:285

[180] Huang L, Jing Y, Chen H, Zhang L, Liu Y. A regional wind wave prediction surrogate model based on CNN deep learning network. Applied Ocean Research. 2022;**126**:103287

[181] Aparna SG, D'souza S, Arjun NB. Prediction of daily sea surface temperature using artificial neural networks. International Journal of Remote Sensing. 2018;**39**:4214

[182] Abisoye BO, Sun Y, Zenghui W. A survey of artificial intelligence methods for renewable energy forecasting: Methodologies and insights. Renewable Energy Focus. 2024;**48**:100529

Chapter 9

# Energy-Efficient Deep Learning Training

*Lei Guan, Shaofeng Zhang and Yongle Chen*

## Abstract

Deep learning has evolved into the most important supporting technology for artificial intelligence (AI) and has achieved widespread application across various fields. However, the energy expenditure associated with training deep learning models has become increasingly significant, now representing an undeniable part of global carbon emissions. This chapter mainly focuses on techniques for achieving energy-efficient deep learning training. It first addresses the context of the significant energy consumption associated with training AI models. Then, it specifically focuses on optimization algorithms and parallel training methods—two key technologies for improving the efficiency of deep learning training. Following that, it presents additional supporting technologies that enhance the training efficiency of AI models. Finally, it provides an overview of specific strategies from a macro perspective.

**Keywords:** carbon neutrality, energy, deep learning, training, interdisciplinary

## 1. Introduction

The Paris Agreement's temperature goal requires achieving carbon neutrality by mid-century [1, 2]. Achieving carbon neutrality targets requires reducing carbon emissions in as many ways as possible [3], especially for those industries requiring a large energy cost. Apart from traditional carbon emissions caused by fossil fuels, the high energy consumption by supporting computing (e.g., cloud computing, edge computing, and high-performance computing) has become a nonnegligible factor contributing to global carbon emissions [4–6] as the energy consumption to power modern computing hardware is directly proportional to carbon dioxide emissions [7].

In recent years, numerous deep neural network (DNN) models have achieved remarkable success across diverse domains, with their training imposing a considerable demand on computational resources and substantial energy consumption [7], particularly in the case of training large language models (LLMs). For instance, the training of GPT-3 with 175 billion parameters is estimated to consume approximately 1.3 gigawatt hours of energy and generate 552 metric tons of carbon emissions [8]. The imperative to achieve energy-efficient deep learning training underscores the critical need to diminish energy consumption and advance toward the carbon neutrality objective.

We assume that $\#\text{epoch}$ is required to train a DNN model to the target accuracy. Letting $t_i$ denote the training duration for the $i$ -th epoch and $t_{\text{total}}$ represent the whole training time, the total training time of DNN models can be formulated as

$$t_{\text{total}} = \sum_{i=1}^{\#\text{epoch}} t_i \approx \#\text{epoch} \times \bar{t}, \tag{1}$$

where $\bar{t}$ denotes the averaged training time, that is, $\bar{t} = \frac{1}{\#\text{epoch}} \sum_{i=1}^{\#\text{epoch}} t_i$ . Eq. (1) reveals that the whole training time of a DNN model is determined by both the convergence, represented by parameter $\#\text{epoch}$ indicating the speed at which the model converges, and the iteration speed, represented by parameter $\bar{t}$ that indicates the speed at which iterations are performed [9]. Therefore, accelerating deep learning model training and reducing energy consumption can be approached by improving both convergence and iteration speed.

A DNN model usually consists of many neural layers, and training the DNN model involves learning the parameters of each layer through an optimization method. The choice of the optimization algorithm is crucial for deep learning training, as different optimization algorithms have varying convergence speeds and, as shown in Eq. (1), can result in different numbers of epochs (i.e., $\#\text{epoch}$ ) used to train the DNN model. More importantly, the optimization algorithm used in deep learning training also determines the accuracy of the resulting model. Developing efficient optimization algorithms often helps reduce the number of epochs required to train a model, which in turn shortens training time and lowers energy consumption for training AI models while using the same computational resources.

Notably, in the era of AI, training deep neural networks often enjoys the following two important characteristics. First, a DNN model is always required to be fed with an immense amount of training data, resulting in quite lengthy and sometimes unacceptable training time. Second, the scale of DNN model parameters has grown dramatically over the years, giving rise to deep learning models with tens to hundreds of layers, totaling billions and even trillions of parameters. In such cases, parallel training often becomes the only viable option for effectively training DNN models. Since parallel training of DNN models often requires coordinating dozens, hundreds, or even larger numbers of GPUs, the energy consumption can be extremely high. As a result, developing efficient parallel training algorithms for DNN models is crucial for enhancing training efficiency and reducing energy consumption.

In the rest of the chapter, we will respectively introduce the main ideas and technologies for advancing the optimization algorithms and parallel training modes associated with deep learning training. Furthermore, we will present additional supporting technologies that contribute to energy-efficient deep learning training. Lastly, we will discuss specific strategies for achieving energy-efficient deep learning training from a macro perspective.

## 2. Optimization algorithms

The training of DNN models involves continuously adjusting the model parameters to find the optimal parameters using an optimizer. First-order gradient methods are currently of core practical importance in deep learning training. During each iteration, a mini-batch training generally consists of one forward pass and one backward propagation, where the gradients with respect to all the parameters (also known as weights) are computed during the backward propagation. After that, the generated gradients are then utilized by the optimizer to calculate the update values for all parameters, which are finally applied to updating the DNN weights [10]. When selecting an optimizer to train a DNN model to an acceptable state, the training time required can vary with different optimizers, even when using the same computational resources. As a result, the energy consumption also differs. Therefore, developing efficient optimizers is crucial for improving the training efficiency of deep learning models and reducing energy consumption.

We use supervised learning to illustrate the application of optimization algorithms. For first-order stochastic optimization methods, the cost function can be written as

$$\min_{\theta \in \mathbb{R}^d} F(\theta) = \frac{1}{n} \sum_{i=1}^{n} \ell\left(y_i, f_\theta(\mathbf{x}_i)\right), \tag{2}$$

where $\ell$ is the loss function, $f_\theta(\mathbf{x}_i)$ is the predicted output with input $\mathbf{x}_i$, and $y_i$ is the target output. We consider the case when training the DNN models using gradient-based optimization methods with a mini-batch size of $b$. At the $t$-th iteration, the gradients in the backward propagation are calculated by

$$\mathbf{g}_t = \frac{1}{b} \sum_{i=1}^{b} \nabla_\theta \ell\left(y_i, f_\theta(\mathbf{x}_i)\right). \tag{3}$$

First-order stochastic optimization algorithms update the weights using the gradients generated by Eq. (3). The difference between various optimization algorithms lies in the different ways they utilize to update weights. The first-order stochastic optimization methods are typically divided into two types: One is the accelerated schemes represented by stochastic gradient descent (SGD), and the other is adaptive methods represented by adaptive moment estimation (Adam) [11].

SGD is a typical optimization algorithm in deep learning. Its approach involves randomly selecting a mini-batch of data samples during each iteration of model updates, calculating their gradients, and then using these gradient values to update the model parameters. For SGD, the parameter update formula goes as

$$\begin{aligned} \mathbf{v}_t &= u\mathbf{v}_{t-1} + (1-\tau)\mathbf{g}_t, \\ \theta_t &= \theta_{t-1} - \gamma \mathbf{v}_t, \end{aligned} \tag{4}$$

where $u$ is the momentum factor, $\tau$ is the dampening for momentum, and $\gamma$ is the learning rate.

For Adam, the parameter update rule is

$$\theta_t = \theta_{t-1} - \frac{\gamma \hat{\mathbf{m}}_t}{\sqrt{\hat{\mathbf{v}}_t} + \epsilon},$$
$$\text{s.t.} \begin{cases} \mathbf{g}_t = \nabla_\theta f_t(\theta_{t-1}), \\ \mathbf{m}_t = \beta_1 \cdot \mathbf{m}_{t-1} + (1-\beta_1) \cdot \mathbf{g}_t, \\ \mathbf{v}_t = \beta_2 \cdot \mathbf{v}_{t-1} + (1-\beta_2) \cdot \mathbf{g}_t^2, \\ \hat{\mathbf{m}}_t = \dfrac{\mathbf{m}_t}{1-\beta_1^t}, \\ \hat{\mathbf{v}}_t = \dfrac{\mathbf{v}_t}{1-\beta_2^t}. \end{cases} \tag{5}$$

In Eq. (5), $\mathbf{m}_t$ and $\mathbf{v}_t$ respectively refer to the exponential moving average (EMA) of $\mathbf{g}_t$ and $\mathbf{g}_t^2$, $\beta_1$ and $\beta_2$ are coefficients used for computing $\mathbf{m}_t$ and $\mathbf{v}_t$ respectively, $\epsilon$ is the smoothing term that can prevent division by zero.

Generally, SGD works well for image classification tasks while Adam and its variants have become the default choice for training many DNN models, particularly the LLMs. Looking at the current development trends in optimization algorithms, adaptive methods should align with the main trends in deep learning optimization. To enhance the effectiveness of optimization algorithms, many methods based on the Adam optimizer have been proposed. The development of optimizers mainly focuses on three aspects: convergence and generalization, computation efficiency, and storage efficiency. These factors collectively determine the effectiveness of optimizers in learning DNN model parameters, thereby impacting energy-efficient deep learning training.

### 2.1 Convergence and generalization

Training DNN models involves repeatedly iterating and updating the model parameters until they converge to the optimal values. Therefore, improving the convergence and generalization of optimization algorithms is crucial for reducing the training time of models and lowering the energy consumption during training.

SGD with momentum (SGDM) accelerates SGD by introducing momentum during the gradient descent process to suppress the oscillations typically seen in SGD. SGDM and SGD share a common feature: They use the same learning rate to update every parameter in the neural network. Notably, neural network models often contain a large number of parameters, which are updated at different frequencies. Adaptive gradient methods, such as Adam, use the EMA of the squared gradients to estimate the learning rate for each parameter. This enables the optimization algorithm to adjust update speeds for different parameters, achieving better convergence than SGDM. To further improve the convergence of adaptive methods, incorporating Nesterov's accelerated gradient (NAG) is a popular choice. Successful instances include Nadam [12], Win [13], NRMSProp [14], AdaPlus [15], and Adan [16]. Furthermore, the recently proposed optimizer Sophia [17] uses a diagonal Hessian-based preconditioner to adapt to the heterogeneous curvatures and improve the convergence.

On the other hand, it has been empirically suggested that the generalization ability of adaptive optimizers is inferior to SGDM on some deep learning tasks such as image classification. To address this issue, many variants of Adam have been proposed, with AdamW [18] being the most notable. AdamW uses $L_2$ regularization or weight decay regularization to train neural networks with Adam to improve Adam's generalization performance. Another successful attempt to improve Adam's generalization ability is AdaBelief [19], which views the EMA of the noisy gradient as the "belief" in the current gradient direction based on which it adjusts the stepsize dynamically based on this belief. Compared to shape minima, flat minima usually lead to better generalization. Sharpness-aware minimization (SAM) can flatten the landscape of deep learning models by introducing extra perturbation steps, thus helping to generalize better when integrated with first-order gradient-based optimizers. Furthermore, weight prediction [10] is also an efficient approach to improve the generalization of gradient-based adaptive methods, including SGDM, Adam, and its variants. Moreover, typical adaptive optimization methods usually suffer from extreme learning rates that can also negatively impact the generalization ability of adaptive optimization algorithms. An efficient solution to this issue is to limit the learning rates within a convergence range bound [20, 21]. A similar approach is Yogi [22], which addresses the nonconvergence issues by controlling the increase in the effective learning rate.

In distributed deep learning training, large batch stochastic optimization methods are employed to train DNN models on massive datasets, which is always computationally challenging. However, large batch training often results in worse convergence and generalization ability. To tackle this problem, a series of methods to enhance the scalability of stochastic optimization algorithms have been proposed. Among them, the most famous approaches are Lars [23] and Lamb [24], the layerwise adaptive large batch optimization technique.

### 2.2 Computation

In each iteration, reducing the computational cost of updating model parameters helps accelerate model training while using the same computational resources, thereby reducing the total energy consumption of model training. As shown in Eq. (5), the main computational cost associated with weight updates consists of the computation of gradients and the computation of first- and second-order moments, with gradients generated after the backward propagation.

There are two main approaches to reducing the computational cost of optimization algorithms. The first approach is to use gradient-free optimization methods to eliminate the computational cost of gradient calculations. A typical example is zero-order optimization that eliminates the need for gradient computation in first-order optimization algorithms, thereby significantly reducing the computational cost per iteration. This is especially relevant when fine-tuning pretrained LLMs, where first-order optimizers such as SGD and Adam require substantial computation overhead for backward propagation. Applying zero-order optimization helps significantly reduce the computational cost for LLM fine-tuning [25].

The second approach is to eliminate the second-order moment calculation in Eq. (5). For instance, Lion [26] optimizer removes the computationally expensive operations including division and square root calculations, making it faster compared to Adam and AdamW.

### 2.3 Memory consumption

The prerequisite for successfully training a DNN model is that both the model parameters and the mini-batch data for training must fit into the GPU memory without causing out-of-memory (OOM) problems throughout the training process. In addition to model parameters and mini-batch data, a GPU is required to maintain optimizer states (e.g., momentum and variances in Adam), gradients, and intermediate activations throughout the entire deep learning training period. Reducing the storage consumption of training deep learning models helps to lower the usage of computational and storage resources, ultimately reducing the energy consumption of training AI models.

Currently, there are two main methods to reduce the memory consumption of optimization algorithms. The first approach is the application of zero-order optimization, which substantially reduces the memory overhead caused by the backward propagation of first-order optimizers. Recent research demonstrates that zero-order optimization helps achieve much better memory efficiency in LLM fine-tuning [25].

The second approach involves removing the second-order moment calculation in classical adaptive methods such as Adam. Classical adaptive optimization methods, including Adam and AdamW, keep track of the second-order moments throughout the training period, which consumes memory equal to the number of parameters. The Lion [26] optimizer is a notable example of a more memory-efficient method, as it only maintains the first-order moment in GPU memory. Additionally, to further reduce the memory consumption of neural network weight matrices, Adafactor [27] proposes maintaining the per-row and per-column sums of these moving averages and then computing the second moment of each parameter based on these sums.

## 3. Parallel training

The essence of parallel training deep neural network models is to use multiple computing devices (usually GPUs) to collaboratively and simultaneously train deep learning models. The training process generally involves deploying the training data or model across multiple computing devices and then accelerating the training process through collaborative and parallel effort. In parallel training deep learning models, the first step is to distribute the data across multiple computing devices. Depending on the way of data distribution, parallel training can generally be categorized into three modes: data parallelism, model parallelism, and hybrid parallelism.

### 3.1 Data parallelism

Data parallelism (DP) is currently the most popular method for parallel training of DNN models. Many popular deep learning frameworks, such as TensorFlow [28] and PyTorch [29], provide convenient APIs to support data parallelism. In data parallelism, each computing device maintains a complete and consistent set of model parameters. These devices are then assigned different batches of training data. After computing gradients, the devices aggregate these gradients through collective communication operations, such as a parameter server or AllReduce, to update the model parameters. It is important to note that the communication required for transmitting gradients is proportional to the size of the entire model parameters. Therefore, as the scale of model parameters increases, the communication overhead for data parallelism also rises. In data parallelism, this communication overhead often becomes a significant factor affecting the training speed and scalability of parallel algorithms. For data parallelism, AllReduce and parameter server are the two most commonly used key technologies, and they directly impact the performance and scalability of data parallelism. In this chapter, we introduce three important approaches to improve the efficiency of data parallelism: the first is selecting an appropriate communication topology, the second is optimizing data communication, and the last is using the Zero Redundancy Optimizer (ZeRO) [30].

Different communication topologies result in varying communication overheads. As shown in **Figure 1**, commonly used communication topologies include star, ring, and tree topologies. The parameter server [31] employs a typical star communication topology. In this topology, the central node is the parameter server node, which stores all the model parameters, while the remaining nodes are worker nodes responsible for performing forward and backward propagation computations. In 2018, Baidu introduced a deep learning parallel training method called Ring-AllReduce, which uses a ring communication topology. As shown in **Figure 1(b)**, each worker node is arranged

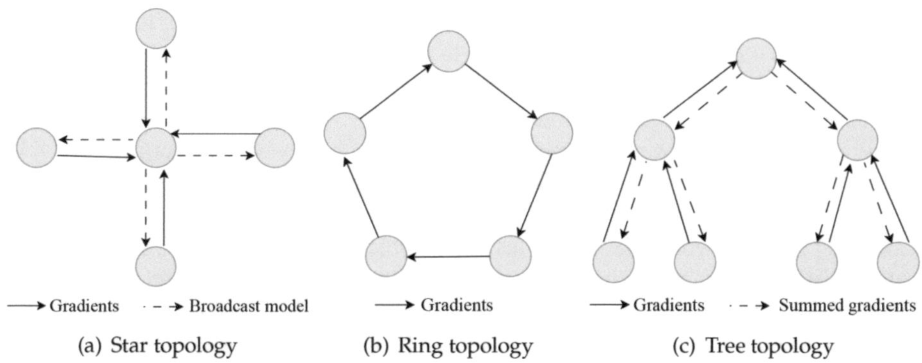


**Figure 1.**
*Three communication topologies for data parallelism.*

in a logical ring with neighboring worker nodes on both its left and right sides. Each worker node receives gradient data from the node on its right and sends the gradients to the node on its left. The Ring-AllReduce has been successfully implemented in the NVIDIA Collective Communications Library (NCCL). The third communication topology is the tree topology, which has also been successfully used in data parallelism [32]. As illustrated in **Figure 1(c)**, the leaf nodes represent all the worker nodes. After each worker node completes backpropagation, the worker nodes collaborate in pairs to compute the sum of their gradients and send the summed gradients to the parent node. This process continues until the root node receives the sum of the gradients from all the leaf nodes. Once the root node has the gradient sum, it broadcasts the sum back to all the leaf nodes. Finally, the leaf nodes perform the parameter update operation.

Optimizing data communication is another crucial approach to improving the efficiency of data parallelism. Common techniques for optimization include reducing communication volume and overlapping communication with computation. Sparse communication is a technique to reduce the communication volume by mapping the majority of small values to zero and communicating only sparse matrices [33]. This allows distributed SGD to proceed more quickly by transmitting sparse updates rather than dense ones. The overlapping technique involves combining communication and computation to enhance efficiency. By utilizing computing resources more effectively, it reduces idle time and accelerates the training process. This approach enables the simultaneous execution of tasks such as gradient calculation or model updates alongside communication activities like gradient aggregation or parameter synchronization.

ZeRO [30] enhances the training efficiency of data parallelism by eliminating memory redundancies while retaining low communication volume and high computational granularity. The key feature of ZeRO is its ability to address OOM issues encountered in data parallelism when training large models by optimizing memory usage. Optimization techniques related to ZeRO include managing model state memory (which includes optimizer states, gradients, and parameters) and optimizing residual state memory (which covers activations, temporary buffers, and unusable memory fragments).

### 3.2 Model parallelism

Model parallelism involves partitioning the neural network model itself, with each computing device receiving a portion of the model. This approach not only overcomes the storage limitations of a single GPU but also results in significantly lower communication overhead compared to data parallelism. Model parallelism can be categorized based on how the DNN models are partitioned into two main types: tensor model parallelism (TMP) and pipeline model parallelism (PMP).

TMP, also known as intralayer model parallelism, involves partitioning the DNN model across GPUs to address the storage limitations of a single GPU. In TMP, the data flow of different operators, such as fully connected layers and convolutional layers, is horizontally partitioned. TMP incurs less communication overhead compared to DP because only a portion of the model parameters needs to be transmitted. However, communication overhead can still be a bottleneck in TMP due to the extensive AllReduce operations required per iteration. PMP is also recognized as interlayer model parallelism, which splits the DNN model into consecutive stages, each consisting of several layers. Once all stages are loaded onto their corresponding GPUs, parallel training is performed in a pipeline manner. During the process of pipeline parallel training, only activation values during forward propagation and gradient values during backward propagation need to be transmitted between neighboring GPUs, resulting in lower communication overhead compared to TMP. Consequently, PMP has become a more popular choice for training DNN models, especially for the training of models with a large number of parameters. Generally, based on the ways of model parameter updates, PMP can be further classified into synchronous PMP and asynchronous PMP. Asynchronous PMP cannot guarantee the same parameter update semantics as data parallelism, often leading to worse convergence and potential accuracy drops. Therefore, synchronous PMP is more commonly used in practical large-model training. Due to space limitations, this chapter will focus on discussing the main methods and techniques for improving the training efficiency of synchronous PMP. A comprehensive review of PMP can be found in Ref. [9].

The most well-known and representative synchronous PMP approach is GPipe [34], which was introduced by Google in 2019. As illustrated in **Figure 2**, GPipe splits a mini-batch into finer-grained micro-batches. During each iteration, these micro-batches are simultaneously fed into the pipeline for pipeline parallel training. GPipe performs synchronous weight updates periodically but suffers from bubble overhead as the pipeline flushes each iteration. In GPipe's pipeline parallel training, "bubbles" represent idle and wasted computational resources. Reducing the number of bubbles in the pipeline can significantly improve training efficiency. Current improvements to GPipe primarily focus on adjusting the scheduling of micro-batches within the pipeline to reduce bubble overhead. **Table 1** summarizes four representative synchronous PMP approaches and the technological means they use to mitigate bubble overhead. However, reducing bubble overhead does not come for free; it often leads to increased computational, communication, and storage overhead. For example, the

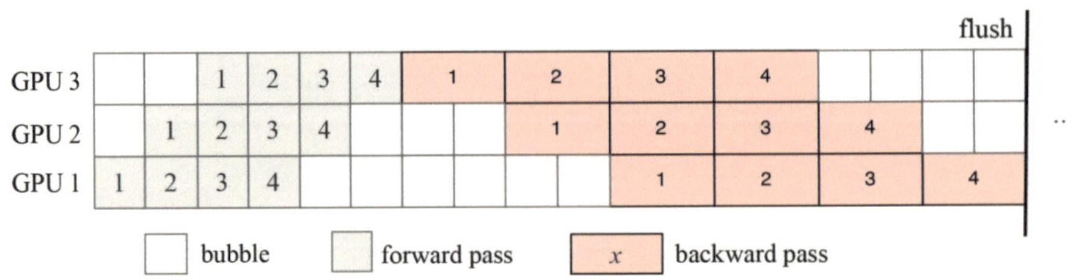


**Figure 2.**
*Illustration of GPipe on 3-GPU computing system. A mini-batch is split into four micro-batches.*

| Synchronous PMP approaches | Technological means |
|---|---|
| DAPPLE [35] | Early backpropagation |
| Chimera [36] | Dual-pipeline structure |
| Megatron-LM [37] | Interleaved "1F1B" schedule |
| Zero Bubble Pipeline [38] | Splitting backward computation |

**Table 1.**
*The four typical synchronous PMP approaches and the used technological means for reducing bubble overhead.*

dual-pipeline structure technique used by Chimera [36] addresses bubbles by adding more pipeline units, but this approach doubles the weight storage overhead and incurs additional communication overhead. Therefore, an optimal PMP approach must strike a balance between minimizing bubble overhead, storage overhead, and communication overhead to achieve the highest training efficiency.

### 3.3 Hybrid parallelism

Hybrid parallelism [39] refers to the simultaneous use of two or more parallelism modes to leverage their advantages for facilitating DNN training. The most popular form of hybrid parallelism is 3D parallelism, which involves combining DP, TMP, and PMP. 3D parallelism integrates the benefits of these three different parallel training modes, offering fast training speed, strong scalability, and high utilization of computational and storage resources. It has become the most important tool for training LLMs that require massive computations. The popular deep learning frameworks, such as Megatron-LM [37], Colossal-AI [40], and DeepSpeed, all support 3D parallelism. **Figure 3** illustrates a 3D parallelism approach used for training BLOOM [41]. DP replicates the model eight times, with each replica placed on a different device and fed a different batch of data. In 12-way PMP, the model layers are partitioned into 12 stages and a stage, with each stage placed on a different GPU. On each device, DP

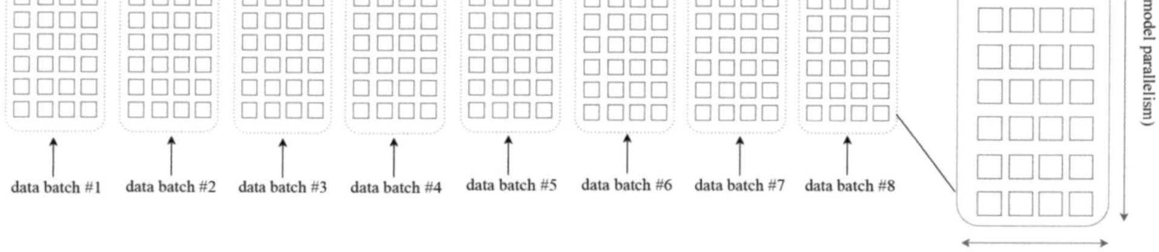


**Figure 3.**
*Illustration of 3D parallelism [41] training on 384 (8 × 12 × 4) GPUs with 8-way DP, 12-way PMP, and 4-way TMP.*

splits each individual layer of the model across four GPUs, so that shards of activation or gradient tensor reside on a single GPU.

Hybrid parallelism is a powerful tool for training large models, especially LLMs. However, it consumes a vast amount of computational and storage resources, leading to significant energy consumption. Therefore, optimizing hybrid parallelism can often result in a substantial reduction in energy consumption. The goal of optimization is to enhance the efficiency and performance of large-scale computations, particularly in large-scale distributed computing systems. The main approaches to optimizing hybrid parallelism involve computation, storage, and communication. Optimizing communication primarily focuses on minimizing interactions between nodes and between GPUs within a node to avoid communication bottlenecks and improve scalability. The communication compression techniques are efficient ways to achieve this goal. Additionally, communication overhead can be mitigated by using overlapping techniques. In terms of storage optimization, ZeRO [30] is a popular choice to enhance storage resource utilization when training LLMs with hybrid parallelism [41]. A similar approach, Zero-Offload [42], moves the tensors from GPU to CPU or NVMe disks to free up space for storing a larger number of model parameters. For computational optimization, hybrid parallelism requires the dynamic distribution of computational tasks across various nodes to reduce idle time and maximize resource utilization.

## 4. Other supporting techniques

### 4.1 Model architecture optimization

Designing energy-efficient AI models is crucial for reducing the environmental impact of AI. Researchers focus on reducing the computational resources required during training and inference, which directly translates to lower electricity usage, reduced carbon emissions, and often faster processing times. Designing efficient AI models benefits both the environment and users, as it reduces the consumption of natural resources and enhances user experience. The model architecture optimization mainly includes the following techniques:

- *Smaller models*: Use smaller, more efficient architectures like MobileNet [43], SqueezeNet [44], or TinyBERT [45], which are specifically designed to reduce computational requirements without significantly compromising performance.

- *Pruning*: Remove unnecessary neurons or layers in a model after training to reduce the size and computation needed during inference. LLM-Pruner is a structural pruning method for LLMs that selectively remove noncritical components based on gradient information to reduce model size while preserving its multitasking and language generation capabilities [46].

- *Quantization*: Reduce the precision of the numbers used in a model, such as using 8-bit integers instead of 32-bit floating-point numbers, to decrease both memory usage and computation [47].

- *Knowledge distillation*: Train a smaller model (student) to replicate the behavior of a larger model (teacher), which can significantly reduce the size and computational cost of the student model [48].

### 4.2 Algorithmic innovations

Algorithmic innovation is also crucial for reducing the energy expenditure of models. Moving forward, we will discuss in detail how these strategies are specifically applied to the transformer model and how they achieve this goal by reducing the amount of computation and the number of operations.

- *Efficient algorithms*: Develop or adopt algorithms that inherently require less computation. For example, attention mechanisms can be optimized to scale better with input size.

- *Sparse representations*: Leverage sparsity in data or model parameters to reduce the number of operations required, which can lead to significant energy savings. Researchers have undertaken a series of efforts to reduce model overhead and thereby improve computational efficiency, such as a range of optimizations for the transformer model. The transformer is a deep learning architecture designed for processing sequential data, first introduced by Vaswani et al. in the 2017 paper "Attention Is All You Need [49]". Sparse transformers [50] are an advanced neural network architecture that addresses the quadratic growth of computation and memory requirements in traditional transformers by employing sparse attention mechanisms, reducing complexity and enabling efficient processing of long sequences.

### 4.3 Efficient training techniques

In addition to the optimization algorithms and parallel training techniques discussed above, there are also several deep learning technologies that can be applied to improve model training efficiency and subsequently reduce energy consumption. Transfer learning, low-rank factorization, and mixed precision training are three prevalent technical methodologies.

- *Transfer learning*: Utilizes pretrained models and fine-tune them on specific tasks rather than training from scratch, which reduces the amount of training time and energy required [51].

- *Low-rank factorization*: Decomposes the weight matrices in neural networks into lower-rank approximations to reduce the number of parameters and operations [52].

- *Mixed precision training*: Uses a combination of lower-precision (e.g., FP16) and higher-precision (e.g., FP32) arithmetic during training to reduce computation without sacrificing model accuracy [53].

### 4.4 Data efficiency

Data, algorithms, and computing power are the three pillars of AI. Techniques related to data efficiency also have a significant impact on the effective training of models. Key data technologies that contribute to this include data augmentation and active learning.

- *Data augmentation*: Uses techniques like data augmentation to increase the diversity of training data without actually increasing the dataset size, which can improve model performance without needing more computational resources [54].

- *Active learning*: Prioritizes the most informative data points during training, reducing the amount of data and computation needed to achieve high performance [55].

## 5. Discussions

The training of AI models (especially LLMs) imposes a considerable demand on energy consumption, running counter to the global goal of carbon neutrality goal by mid-century [56]. In addressing this imperative, AI researchers and practitioners are urged to adopt specific measures for realizing energy-efficient deep learning training. Training deep learning models involves knowledge and techniques from multiple fields, including machine learning, optimization algorithms, parallel and distributed algorithms, systems, and computer architecture. In general, as shown in **Figure 4**, it requires efforts and techniques at the following four levels.

- *Hardware level*: Accelerators, represented by GPUs, are now widely used in training AI models. Priority should be given to the hardware level, particularly the design of energy-efficient AI chips [57].

- *System level*: Increased attention should be directed toward data storage systems, as both model parameters and data scale continue to grow rapidly in the era of large models. Furthermore, implementing efficient job scheduling helps optimize the use of computational resources, prevents computing resource wastage, and minimizes unnecessary carbon emissions as much as possible. Lastly, the popular deep learning training frameworks such as PyTorch and TensorFlow provide systematic support for training deep learning models. Performing continuous optimization and improvement of deep learning training frameworks are highly beneficial in reducing the energy consumption caused by training AI models.

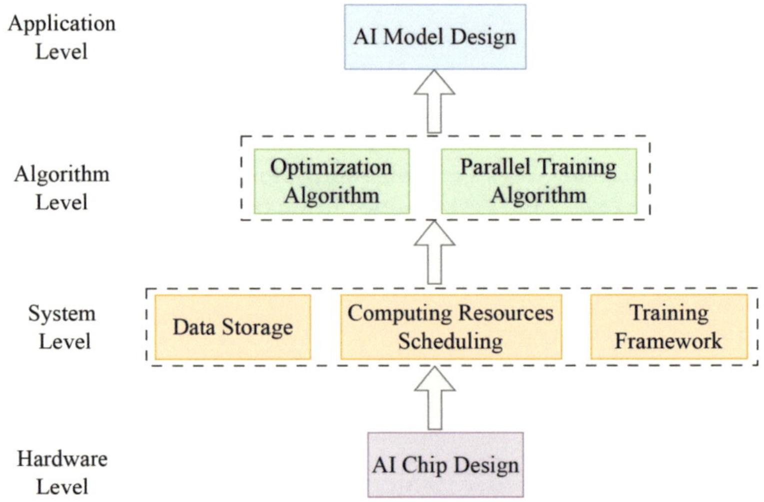


**Figure 4.**
*The framework for achieving energy-efficient deep learning training.*

- *Algorithm level*: Dedicated efforts are required to formulate energy-efficient deep learning training algorithms. This entails developing more efficient parallel algorithms to accelerate iterations and implementing optimization methods to curtail training iterations—both factors jointly determining the training duration of deep learning models.

- *Application level*: Focusing on the development of energy-efficient AI models is essential for reducing the energy consumption involved in training AI models. Researchers should aim to reduce the computational load by minimizing the number of parameters and computational complexity, all while maintaining model performance.

Furthermore, as suggested in Ref. [56], the government should act as the gatekeepers of carbon neutrality, formulating and supervising the effective execution and implementation of various policies. Achieving low-carbon AI calls for both interdisciplinary research and international cooperation, which we refer to 2I policy. The 2I policy calls for ad-hoc research and cooperation between the AI industry and the energy sector.

## 6. Conclusions

Achieving energy-efficient deep learning training helps reduce the energy consumption associated with training deep learning models, thereby supporting the global carbon neutrality strategy. This chapter systematically discusses the key methods and techniques involved in training deep learning models, with the aim of providing valuable insights for reducing carbon emissions from AI model training and promoting carbon neutrality.

## Acknowledgements

This work is supported in part by the National Science Foundation of China under Grant No. 62472302, the State Administration of Science, Technology, and Industry for National Defense of China under Grant No. WDZC20235250118, and the Shanxi Province Special Fund for Guiding the Transformation of Scientific and Technological Achievements under Grant No. 202304021301037.

The authors acknowledge the use of ChatGPT for the language polishing of the manuscript.

## Conflict of interest

The author declares no conflict of interest.

**Author details**

Lei Guan[1,2]*, Shaofeng Zhang[3] and Yongle Chen[1,2]*

1 College of Computer Science and Technology, Taiyuan University of Technology, Taiyuan, China

2 Shanxi Provincial Key Laboratory of Industrial Internet Security, Taiyuan, China

3 Tencent Technology, Beijing, China

*Address all correspondence to: guanleimath@163.com and chenyongle@tyut.edu.cn

## References

[1] Huang M-T, Zhai P-M. Achieving Paris agreement temperature goals requires carbon neutrality by middle century with far-reaching transitions in the whole society. Advances in Climate Change Research. 2021;**12**(2):281-286

[2] Wang F, Harindintwali JD, Yuan Z, Wang M, Wang F, Li S, et al. Technologies and perspectives for achieving carbon neutrality. The Innovation. 2021;**2**(4):1-22

[3] Chen JM. Carbon neutrality: Toward a sustainable future. The Innovation. 2021;**2**(3):1-2

[4] Bharany S, Sharma S, Khalaf OI, Abdulsahib GM, Al Humaimeedy AS, Aldhyani TH, et al. A systematic survey on energy-efficient techniques in sustainable cloud computing. Sustainability. 2022;**14**(10):6256

[5] Iyer SS, Roychowdhury V. Ai computing reaches for the edge. Science. 2023;**382**(6668):263-264

[6] Shalf J. The future of computing beyond moore's law. Philosophical Transactions of the Royal Society A. 2020;**378**(2166):20190061

[7] Strubell E, Ganesh A, McCallum A. Energy and policy considerations for modern deep learning research. Proceedings of the AAAI Conference on Artificial Intelligence. 2020;**34**(09):13 693-13 696

[8] Patterson D, Gonzalez J, Le Q, Liang C, Munguia LM, Rothchild D, et al. Carbon Emissions and Large Neural Network Training. arXiv Preprint arXiv:2104.10350. 2021

[9] Guan L, Li D-S, Liang J-Y, Wang W-J, Ge K-S, Lu X-C. Advances of pipeline model parallelism for deep learning training: An overview. Journal of Computer Science and Technology. 2024;**39**(3):567-584

[10] Guan L, Li D, Shi Y, Meng J. Xgrad: Boosting gradient-based optimizers with weight prediction. IEEE Transactions on Pattern Analysis and Machine Intelligence. 2024;**46**(10):6731-6747

[11] Kingma DP. Adam: A Method for Stochastic Optimization. arXiv preprint arXiv:1412.6980. 2014

[12] Dozat T. Incorporating Nesterov Momentum Into Adam. 2016

[13] Zhou P, Xie X, Yan S. Win: Weight-Decay-Integrated Nesterov Acceleration for Adaptive Gradient Algorithms. 2023

[14] Elshamy R, Abu-Elnasr O, Elhoseny M, Elmougy S. Improving the efficiency of rmsprop optimizer by utilizing nestrove in deep learning. Scientific Reports. 2023;**13**(1):8814

[15] Guan L. Adaplus: Integrating Nesterov momentum and precise stepsize adjustment on adamw basis. In: ICASSP 2024-2024 IEEE International Conference on Acoustics, Speech and Signal Processing (ICASSP). Seoul, Korea: IEEE; 2024. pp. 5210-5214

[16] Xie X, Zhou P, Li H, Lin Z, Yan S. Adan: Adaptive Nesterov momentum algorithm for faster optimizing deep models. IEEE Transactions on Pattern Analysis and Machine Intelligence. 2024

[17] Liu H, Li Z, Hall D, Liang P, Ma T. Sophia: A Scalable Stochastic Second-Order Optimizer for Language Model Pre-Training. arXiv preprint arXiv:2305.14342. 2023

[18] Loshchilov I. Decoupled Weight Decay Regularization. arXiv preprint arXiv:1711.05101. 2017

[19] Zhuang J, Tang T, Ding Y, Tatikonda SC, Dvornek N, Papademetris X, et al. Adabelief optimizer: Adapting stepsizes by the belief in observed gradients. Advances in Neural Information Processing Systems. 2020;**33**:18 795-18 806

[20] Liao X, Sahran S, Abdullah A, Shukor SA. Adacb: An adaptive gradient method with convergence range bound of learning rate. Applied Sciences. 2022;**12**(18):9389

[21] Luo L, Xiong Y, Liu Y, Sun X. Adaptive Gradient Methods with Dynamic Bound of Learning Rate. arXiv preprint arXiv:1902.09843. 2019

[22] Zaheer M, Reddi S, Sachan D, Kale S, Kumar S. Adaptive methods for nonconvex optimization. Advances in Neural Information Processing Systems. 2018;**31**:9815-9825

[23] You Y, Gitman I, Ginsburg B. Large Batch Training of Convolutional Networks. arXiv preprint arXiv:1708.03888. 2017

[24] You Y, Li J, Reddi S, Hseu J, Kumar S, Bhojanapalli S. Large Batch Optimization for Deep Learning: Training Bert in 76 Minutes. arXiv preprint arXiv:1904.00962. 2019

[25] Zhang Y, Li P, Hong J, Li J, Zhang Y, Zheng W, et al. Revisiting Zeroth-Order Optimization for Memory-Efficient llm Fine-Tuning: A Benchmark. arXiv preprint arXiv:2402.11592. 2024

[26] Chen X, Liang C, Huang D, Real E, Wang K, Pham H, et al. Symbolic discovery of optimization algorithms. Advances in Neural Information Processing Systems. 2024;**36**:49205-49233

[27] Shazeer N, Stern M. Adafactor: Adaptive learning rates with sublinear memory cost. In: International Conference on Machine Learning. Stockholm, Sweden: PMLR; 2018. pp. 4596-4604

[28] Abadi M, Barham P, Chen J, Chen Z, Davis A, Dean J, et al. Tensorflow: A system for large-scale machine learning. In: 12th USENIX Symposium on Operating Systems Design and Implementation (OSDI 16). Savannah, GA, USA: USENIX Association; 2016. pp. 265-283

[29] Paszke A, Gross S, Massa F, Lerer A, Bradbury J, Chanan G, et al. Pytorch: An imperative style, high-performance deep learning library. Advances in Neural Information Processing Systems. 2019;**32**:8026-8037

[30] Rajbhandari S, Rasley J, Ruwase O, He Y. Zero: Memory optimizations toward training trillion parameter models. In: SC20: International Conference for High Performance Computing, Networking, Storage and Analysis. Atlanta, GA, USA: IEEE; 2020. pp. 1-16

[31] Li M, Andersen DG, Park JW, Smola AJ, Ahmed A, Josifovski V, et al. Scaling distributed machine learning with the parameter server. In: 11th USENIX Symposium on Operating Systems Design and Implementation (OSDI 14). Broomfield, CO: USENIX Association; 2014. pp. 583-598

[32] Iandola FN, Moskewicz MW, Ashraf K, Keutzer K. Firecaffe: Near-linear acceleration of deep neural network training on compute clusters. In: Proceedings of the IEEE Conference

on Computer Vision and Pattern Recognition. Las Vegas, NV, USA: IEEE; 2016. pp. 2592-2600

[33] Aji AF, Heafield K. Sparse Communication for Distributed Gradient Descent. arXiv preprint arXiv:1704.05021. 2017

[34] Huang Y, Cheng Y, Bapna A, Firat O, Chen D, Chen M, et al. Gpipe: Efficient training of giant neural networks using pipeline parallelism. Advances in Neural Information Processing Systems. 2019;**32**:103-112

[35] Fan S, Rong Y, Meng C, Cao Z, Wang S, Zheng Z, et al. Dapple: A pipelined data parallel approach for training large models. In: Proceedings of the 26th ACM SIGPLAN Symposium on Principles and Practice of Parallel Programming. ACM; 2021. pp. 431-445

[36] Li S, Hoefler T. Chimera: Efficiently training large-scale neural networks with bidirectional pipelines. In: Proceedings of the International Conference for High Performance Computing, Networking, Storage and Analysis. 2021. pp. 1-14

[37] Narayanan D, Shoeybi M, Casper J, LeGresley P, Patwary M, Korthikanti V, et al. Efficient large-scale language model training on gpu clusters using megatron-lm. In: Proceedings of the International Conference for High Performance Computing, Networking, Storage and Analysis. St. Louis, Missouri, USA: ACM; 2021. pp. 1-15

[38] Qi P, Wan X, Huang G, Lin M. Zero bubble (almost) pipeline parallelism. In: The Twelfth International Conference on Learning Representations. 2024

[39] Ben-Nun T, Hoefler T. Demystifying parallel and distributed deep learning: An in-depth concurrency analysis. ACM Computing Surveys (CSUR). Vienna, Austria: International Conference on Learning Representations (ICLR); 2019;**52**(4):1-43

[40] Li S, Liu H, Bian Z, Fang J, Huang H, Liu Y, et al. Colossal-ai: A unified deep learning system for large-scale parallel training. In: Proceedings of the 52nd International Conference on Parallel Processing. Salt Lake City, UT, USA: ACM; 2023. pp. 766-775

[41] Le Scao T, Fan A, Akiki C, Pavlick E, Ilić S, Hesslow D, et al. Bloom: A 176b-Parameter Open-Access Multilingual Language Model. 2023

[42] Ren J, Rajbhandari S, Aminabadi RY, Ruwase O, Yang S, Zhang M, et al. Zero-offload: Democratizing billion-scale model training. In: 2021 USENIX Annual Technical Conference (USENIX ATC 21). USENIX Association; 2021. pp. 551-564

[43] Howard AG, Zhu M, Chen B, Kalenichenko D, Wang W, Weyand T, et al. Mobilenets: Efficient Convolutional Neural Networks for Mobile Vision Applications. arXiv preprint arXiv:1704.04861. vol. 126. 2017

[44] Iandola FN. Squeezenet: Alexnet-Level Accuracy with 50x Fewer Parameters and¡ 0.5 mb Model Size. arXiv preprint arXiv:1602.07360. 2016

[45] Jiao X, Yin Y, Shang L, Jiang X, Chen X, Li L, et al. Tinybert: Distilling Bert for Natural Language Understanding. arXiv preprint arXiv:1909.10351. 2019

[46] Ma X, Fang G, Wang X. Llm-pruner: On the structural pruning of large language models. Advances in Neural Information Processing Systems. 2023;**36**:21 702-21 720

[47] Gholami A, Kim S, Dong Z, Yao Z, Mahoney MW, Keutzer K. A survey of quantization methods for efficient neural network inference. In: Low-Power Computer Vision. 1em plus 0.5em minus 0.4em Chapman and Hall/CRC. 2022. pp. 291-326

[48] Huang T, You S, Wang F, Qian C, Xu C. Knowledge distillation from a stronger teacher. Advances in Neural Information Processing Systems. 2022;**35**:33 716-33 727

[49] Vaswani A. Attention is All You Need, Advances in Neural Information Processing Systems. 2017

[50] Child R, Gray S, Radford A, Sutskever I. Generating Long Sequences with Sparse Transformers. arXiv preprint arXiv:1904.10509. 2019

[51] Weiss K, Khoshgoftaar TM, Wang D. A survey of transfer learning. Journal of Big Data. 2016;**3**:1-40

[52] Swaminathan S, Garg D, Kannan R, Andres F. Sparse low rank factorization for deep neural network compression. Neurocomputing. 2020;**398**:185-196

[53] Micikevicius P, Narang S, Alben J, Diamos G, Elsen E, Garcia D, et al. Mixed Precision Training. arXiv preprint arXiv:1710.03740. 2017

[54] Zhong Z, Zheng L, Kang G, Li S, Yang Y. Random erasing data augmentation. Proceedings of the AAAI Conference on Artificial Intelligence. 2020;**34**(07):13 001-13 008

[55] Ren P, Xiao Y, Chang X, Huang P-Y, Li Z, Gupta BB, et al. A survey of deep active learning. ACM Computing Surveys (CSUR). 2021;**54**(9):1-40

[56] Guan L. Reaching carbon neutrality requires energy-efficient training of AI. Nature. 2024;**626**(7997):33-33

[57] Larson C. China's AI Imperative. 2018